The Art of the Cabinet Maker

The Art of the
CABINET MAKER

ROBERT COOKSEY
and PAUL BRAMWELL

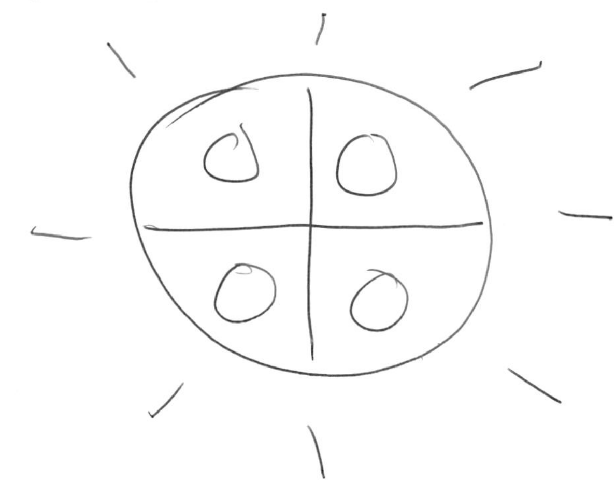

Together with all Nations we protect both Life and Land and hold the Worlds in Balance.

Hopi

The Crowood Press

First Published in 1996 by
The Crowood Press Ltd
Ramsbury, Marlborough
Wiltshire SN8 2HR

British Library Cataloguing-in-Publication Data

A catalogue reference for this book is available from the British Library.

ISBN 1 85223 982 4

Line illustrations by David Fisher.
Photographs by Chris Ashdown.

Typefaces used: text, New Baskerville and Garamond; headings, Optima Bold.

Typeset and designed by
D & N Publishing, Ramsbury, Wiltshire.

Printed in Great Britain by BPC Books, Aylesbury.

CONTENTS

Introduction

It is difficult to go anywhere without coming into contact with wood in one form or another. It constitutes the desks on which we work, the tables at which we eat and the beds in which we sleep. Its usage has dwindled with the advent of synthetic materials and the use of metal. At the same time, the ways in which wood is used have been drastically altered with the introduction of mass production techniques. This has inevitably led to the decline of traditional crafts and the loss of much of the expertise and wisdom that in former times were commonplace.

But things are not always as doom laden as they may at first appear. The cabinet maker's craft – as any other – is influenced by wider changes in fashion and living styles, so it is important to be willing and able to change and adapt to new circumstances. The advantage for the cabinet maker is that there are centuries of tradition and experience on which to draw when these adaptations permeate into the practicalities of the workshop. The present day situation is simply one more step in that long journey.

The journey began in the fifteenth century. Until then – and since the earliest days of communal living – wood had been used for housing, primitive weaponry, agricultural appliances and domestic furniture. However, it was usually home-made and rough hewn, and utilitarian in purpose and design. For all but a very small minority of the population there was little to store and precious little time in which to lounge around the home. Elaborate items of furniture were not only expensive, they were not actually needed.

The turning point came when society became more sophisticated and wealthier. There was the emergence of a merchant middle class who wanted to show off their new found wealth through the construction of larger and more luxurious houses. Crudely cut tables and chairs did not fit well into this image and so for the first time there was a significant demand for well-made furniture on which busy souls could rest their weary limbs after a long day at the exchanges and in which they could store and show off their newly acquired consumables. Out of this demand emerged the skilled cabinet maker and an apprenticeship system that has been perpetuated and has built upon these early foundations.

There were other important factors. The development of science and technology had an impact throughout on the nature and type of tools that could be employed. Improvements in transport allowed the craft to flourish and expand. Over time it is possible to see how tastes and fashions, influences from other countries and the arts have affected not only the items that have been made, but the style in which they were made.

Perhaps the biggest influence of all has been the nature of the housing in which the furniture made by the cabinet maker has to be put. The large houses of the wealthy enabled the construction of large pieces over long periods of time. The abundance of domestic servants meant that cleaning and maintenance of these pieces was not a problem. This is as much an influence upon the present day craft as it ever has been. The grand old houses are now largely gone, at least out of private hands. People have smaller houses, busier lives and have more choice as to how to spend their disposable income. It is against this background that this book and the techniques described in it should be set.

The book is intended to be a general and basic introduction to the techniques employed by the modern cabinet maker. In scope and intent, it is not designed to be exhaustive. (Indeed, this is such a huge topic that it is doubtful if any work can truly cover all that is important in this field.) It is hoped that the most important techniques that have stood the test of time have been identified and have been included – because these are the techniques that are still fundamental to the craft as a whole – while at the same time not shying away from complementary new materials, equipment and techniques. It is hoped that this book will be a starting point in a new adventure for the enthusiast who has some prior knowledge and who wishes to build upon it in a practical and easy-to-read way before going into greater depth and detail at a later stage.

CHAPTER

one

Tools and Equipment

It is difficult to over-emphasize the importance of the selection of the correct tools and equipment before starting even the simplest of jobs. The purchase of incorrect or inappropriate tools and equipment will seriously impair the quality of any job that is undertaken – and it is a false economy to obtain cheap or damaged equipment. Cabinetry is a precise art and it needs precision tools that are inevitably expensive. It is therefore crucial that only the tools that are absolutely necessary are chosen, but that these are chosen wisely and without parsimony. High-quality tools that are carefully maintained will last a lifetime.

THE WORKBENCH

The workbench is the most important piece of equipment for any cabinet maker. There are many stories of individuals producing fine pieces of furniture on the kitchen or dining-room table, but such stories need to be treated with a degree of caution. The workbench is the centre of the cabinet maker's universe, and much of

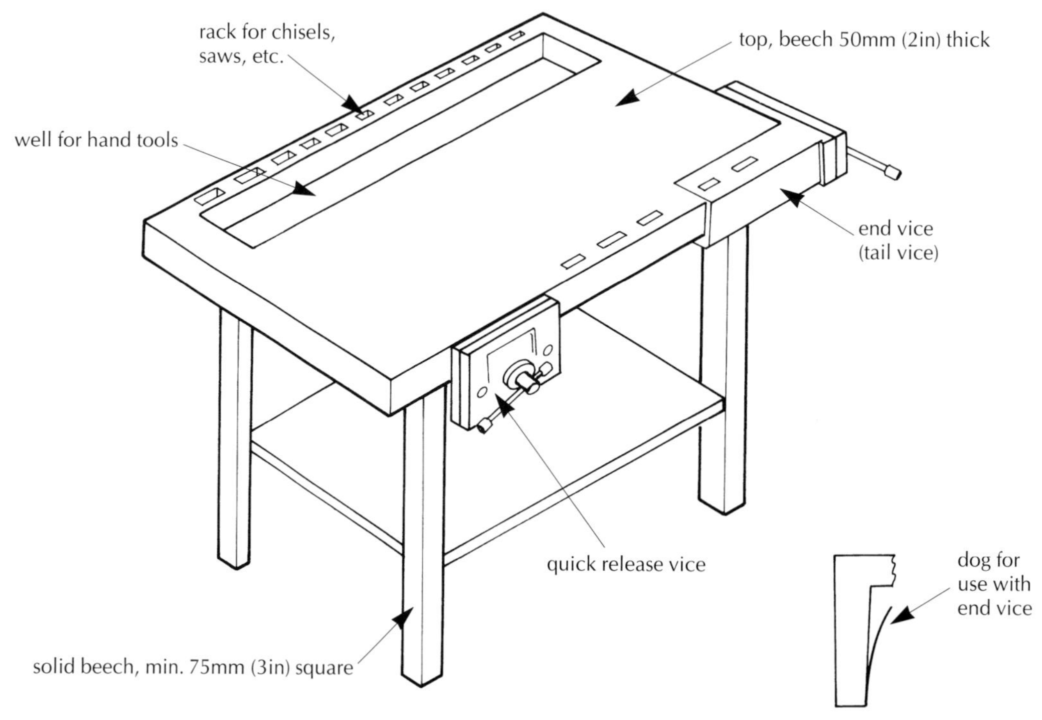

Parts of the workbench.

the work revolves around it. It is therefore critical that it is up to the task. Although there are many excellent fold-up workbenches on the market, their uses in this area are severely limited and it is probably wise to go for a solid traditional workbench.

Ideally this should be made of a sturdy hardwood such as beech. A softwood is not advisable for the top as it is not sufficiently dense and will tend to warp with time. To save money the support frame may be made of a cheaper timber, such as a softwood like deal. Benches can be purchased as complete pieces or can be made and customized to meet individual requirements. However it is acquired, make sure that the top front rail is notched over the legs and screwed in. The long lower rails should be securely held with wedges. Thus the whole thing can be taken apart and moved if need be.

The working surface needs to be large enough to handle quite large items, but not too large that work has to be done at a stretch. It is also important to remember that the larger the working surface the greater the expense in terms of timber.

The next important consideration is the question of what type of vice to fit. There is a wide selection and the choice will depend on the type of work that is to be done. The bigger and heavier the material being worked on, the stronger the vice needs to be. For the person just starting out the most useful accessories would be an end vice and a quick release vice. Start with two vices that are good general models of good quality. When you become clearer in your own mind as to which areas you are going to work in, that is the time to examine more specialist vices.

It is important to plan the spot where the workbench is to be placed. Remember that you will need room to work around the workbench and that you may have large pieces of timber protruding from it. For many people this is a problem when the only space available is a confined garden shed or a garage, but bear in mind the problem of working space whenever possible.

The workbench should be sturdy enough to support heavy weights and the working surface should be perfectly flat and level. A badly aligned working surface will create major difficulties at a later date. One of the things that is commonly overlooked is the foundation upon which the workbench is placed. Examine the workshop floor carefully and make sure that it is solid and flat. Reinforce it with concrete if possible, as it may have to withstand considerable weight.

MARKING OUT TOOLS

Tradition has it that the cabinet maker measures and marks out twice and cuts only once. It is sage advice. There is virtually no job that can be undertaken that does not require accurate marking out, be it marking parallel lines from the edge of a piece of timber, marking points for additional fittings such as handles and locks or precise markings for joints. The consequences of short-cutting this procedure are self evident.

Marking out requires skill and experience, but it also requires the correct application of the most suitable tools for the job. When choosing measuring tools the key concern should be for quality, even if this stretches the budget a little. Always start from the point of view of what is needed to complete a particular task to avoid unnecessary expenditure. However, there is a general nucleus of tools and equipment that it would be difficult to do without.

TRY SQUARE

Foremost among tools in this area should be the selection of a top-class square. Choose a traditional try square which has a

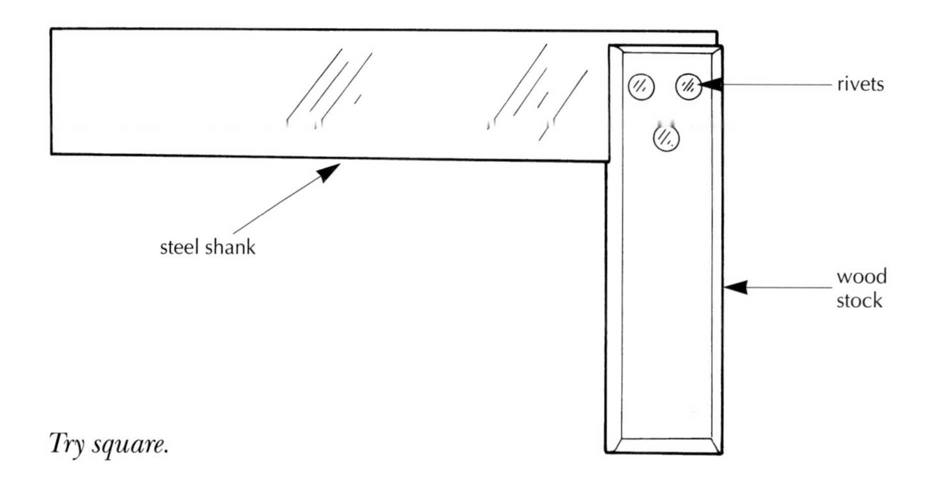

Try square.

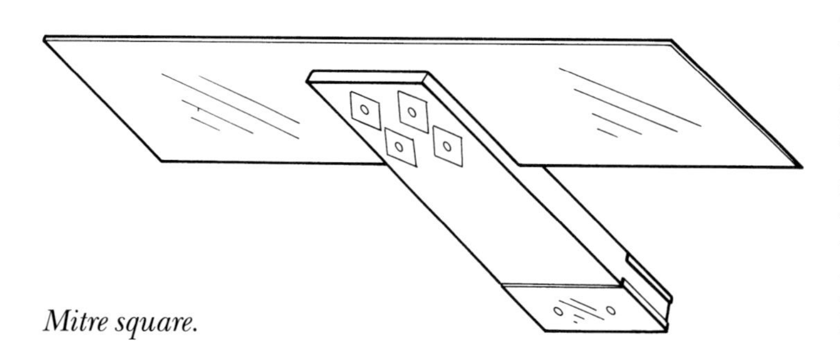

Mitre square.

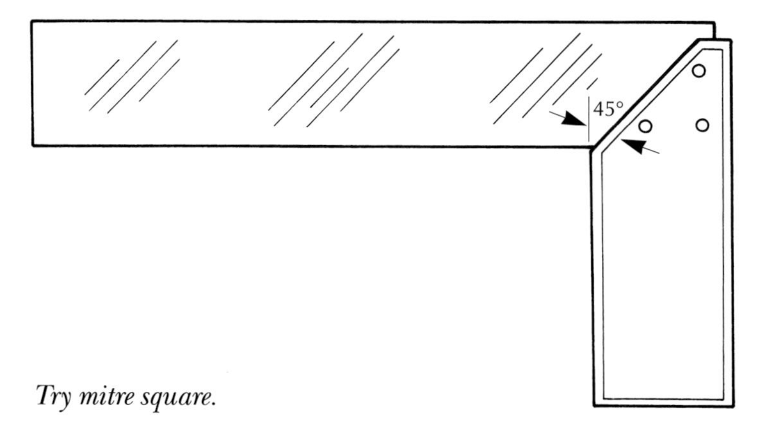

Try mitre square.

hardened steel blade securely fixed with rivets to a wooden stock. The stock should be faced with brass for accuracy and protection. Try squares can be purchased in various sizes and the choice is a personal one, but it would be useful to have a 150mm (6in) and a 300mm (12in) square.

MITRE SQUARE

This has a hardwood stock and a fine metal blade set at an angle of 45°. As its name suggests it is designed for marking and testing mitres. It is possible to obtain an adjustable version of this tool – the blade can be set anywhere between 45° and 90° and it has a metal stock. In most instances it is better to stick to the fixed variety.

TRY MITRE SQUARE

This is similar in appearance to the standard try

square, except that it incorporates an angle of 45° instead of the regular 90°. It is used for speedily measuring 45° angles and generally has a stock made of cast metal rather than the wooden stock of its more familiar cousin.

ENGINEER SQUARE

This is a precision ground tool with hardened and tempered blades. It is a more compact, less cumbersome and better balanced piece of equipment and in many ways easier to use than a conventional try square. A 150mm (6in) engineer square is

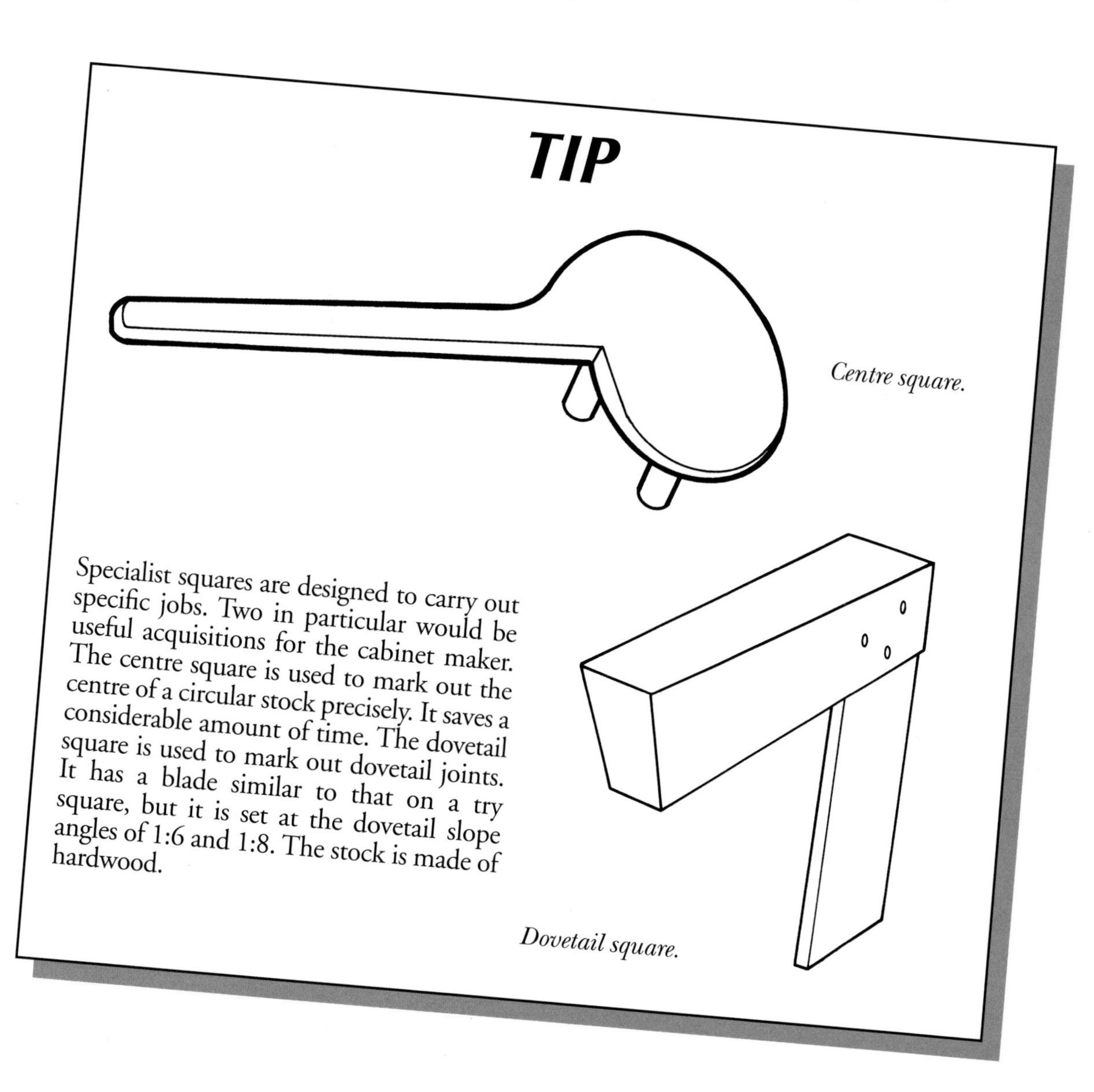

TIP

Centre square.

Specialist squares are designed to carry out specific jobs. Two in particular would be useful acquisitions for the cabinet maker. The centre square is used to mark out the centre of a circular stock precisely. It saves a considerable amount of time. The dovetail square is used to mark out dovetail joints. It has a blade similar to that on a try square, but it is set at the dovetail slope angles of 1:6 and 1:8. The stock is made of hardwood.

Dovetail square.

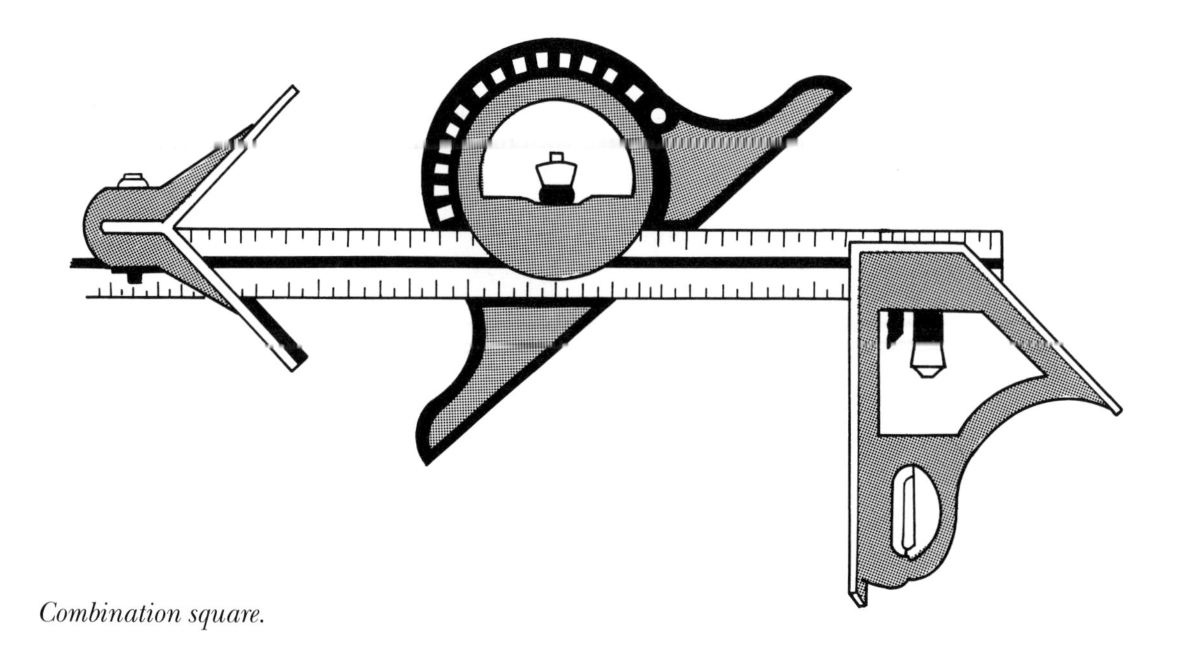

Combination square.

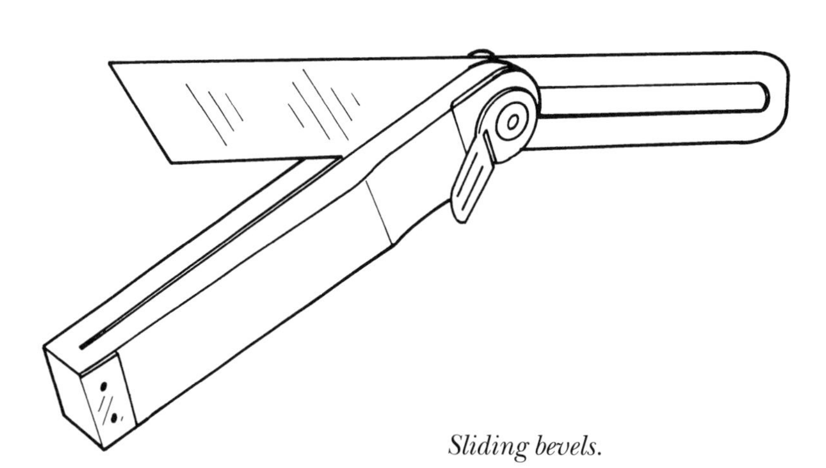

Sliding bevels.

particularly useful if you are working within very fine parameters and it costs little more than a standard try square.

COMBINATION SQUARE

A combination square is a very useful tool that serves a range of purposes, especially if it is fitted with a centre square and a protractor. It has the advantage of being compact and flexible and saves the irritation of having to swap tools in the middle of a job.

SLIDING BEVELS

Sliding bevels are indispensable for measuring all those odd angles that fall between rigid squares. The blade is adjustable to any angle and locked in place either by a slotted screw or a brass locking nut; the stock is generally constructed of hardwood. The sliding bevel is useful for marking out dovetails in the absence of a specialist dovetail square. One point to remember when using this tool is to make sure that the butt is firmly held square to the edge of the timber as failure to do this will result in a false reading.

DOVETAIL TEMPLATES

These consist of pieces of metal folded or excluded to form a right angle. They are cut with a slope of 1:6 for softwood and 1:8 for hardwood. They are helpful in marking out dovetail joints speedily and accurately.

MARKING KNIFE

This is a tool that the cabinet maker simply cannot do without. Although it is sometimes possible to use a chisel, the marking knife is essential when the marking out is fine and will line the timber more accurately than a pencil. There are many varieties available, with little to chose between them, but make sure that the blade is of good quality tempered steel. The handle can be of beech, rosewood or polypropylene.

RULES

The traditional rule of the carpenter is the *hardwood folding rule*. It is 1,000mm (39⅜in) long and has a metric scale. It has brass tips and joints for protection against the inevitable wear it will have. It slips easily into the pocket and is easy to use.

TIP

When using any rule make sure that the markings actually touch the wood to avoid the danger of taking inaccurate readings.

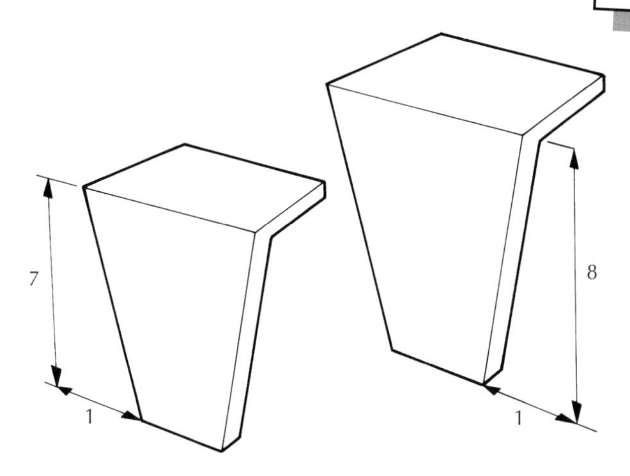

Dovetail templates.

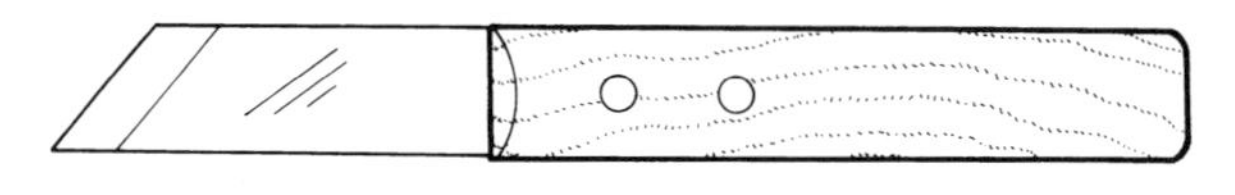

Marking knife.

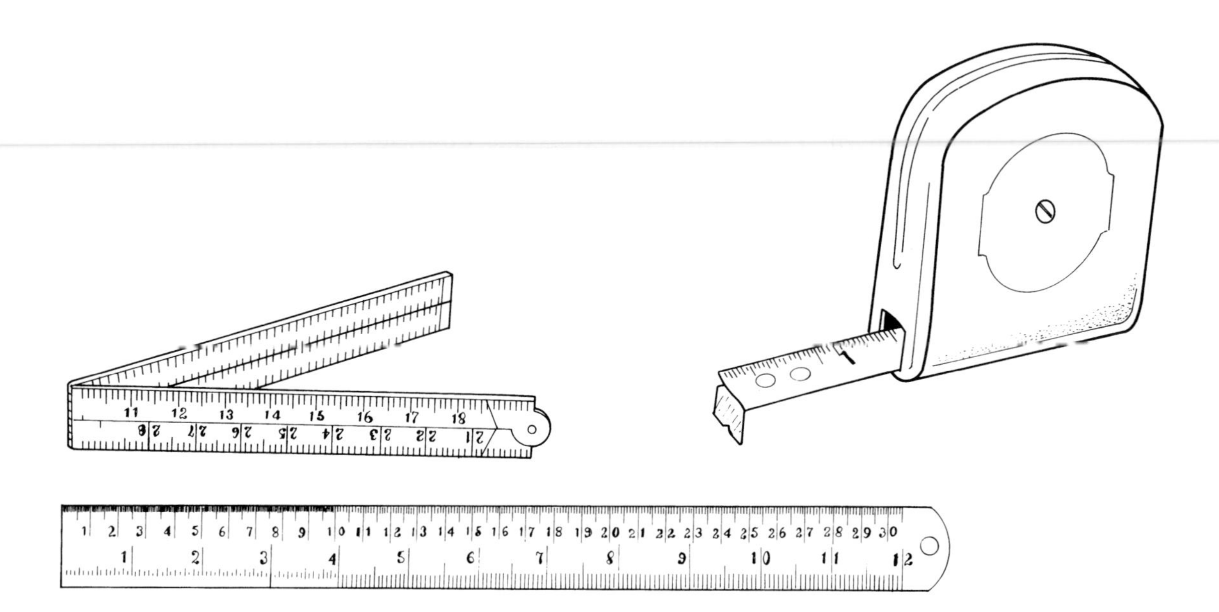

Rules and tape.

Precision steel rules come in a number of sizes ranging from 150 to 1250mm (6 to 49in). They usually come with an easy-to-read satin anti-glare finish, important when working in the artificial light of many workshops.

TAPES

Not everything needs to be measured with a high degree of accuracy, and a common retractable steel tape measure is handy to have around. They have both metric and imperial scales, are cheap and can readily be purchased from any hardware supplier.

GAUGES

Gauges are used for marking lines parallel to the edge of a piece of timber. They are an essential part of a cabinet maker's armoury as they are crucial for marking joints, the thickness and width of timber and so on. As with other marking tools a range of gauges will be needed.

MARKING GAUGE

This is used for marking parallel to an edge with the grain or on an end grain. It is predominantly used for marking width and thickness. There are metal and wood versions available – the metal variety is a little more expensive but has no major advantages over its beech or rosewood counterpart.

CUTTING GAUGE

This gauge can be used for cutting and gauging across the timber and is the same as the marking gauge except that it has a sharp steel blade instead of a spur, held in place by a brass wedge. Although it is intended for cutting thin pieces of wood, the fineness of the blade makes it just as capable of marking timber across the grain.

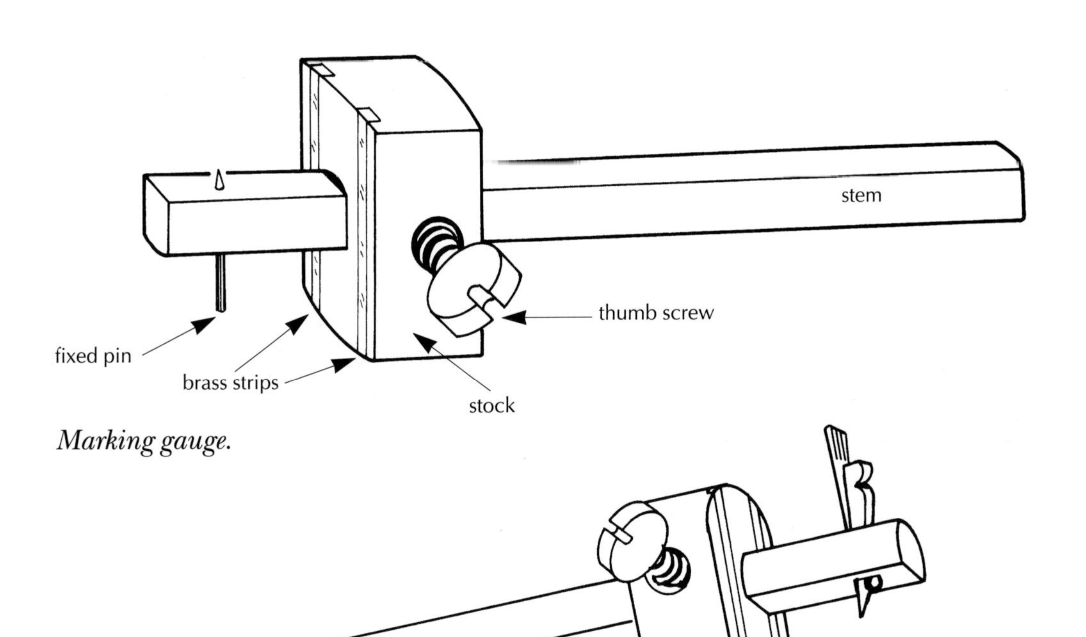

Marking gauge.

Cutting gauge.

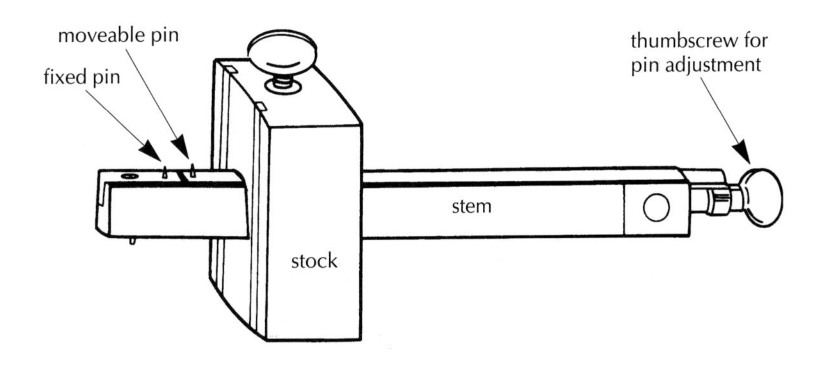

Mortise gauge.

MORTISE GAUGE

This is used for marking out mortise and tenon joints. It can be distinguished from the marking gauge in that it has two spurs instead of one. These are adjustable to allow two lines to be marked parallel to a prepared edge.

> **TIP**
>
> It is important to remember that a marking gauge or a mortise gauge must be used from the face side of the wood.

PLANES

Planes fulfil a range of functions, but primarily they are used for reducing the thickness or width of a piece of timber, producing a smooth surface or for straightening. Specialist planes are used for operations such as grooving, rebating and moulding. Planing techniques are described in Chapter 4.

BENCH PLANES

The bench plane is the work-horse that undertakes the everyday, non-specialist tasks of the cabinet maker. Bench planes come in many sizes and are categorized by their length and cutter width. They used to be made of wood, but they are rapidly being replaced by metal planes that are often easier to use.

The way in which the plane is used is largely determined by its length – the larger the piece of work the longer the plane needs to be to avoid the danger of hollowing. Below is a brief description of a range of the most useful bench planes, starting with the smallest.

Smoothing Plane

Various types are available, but as the name suggests they are all used to smooth the surface on the final cut and add the

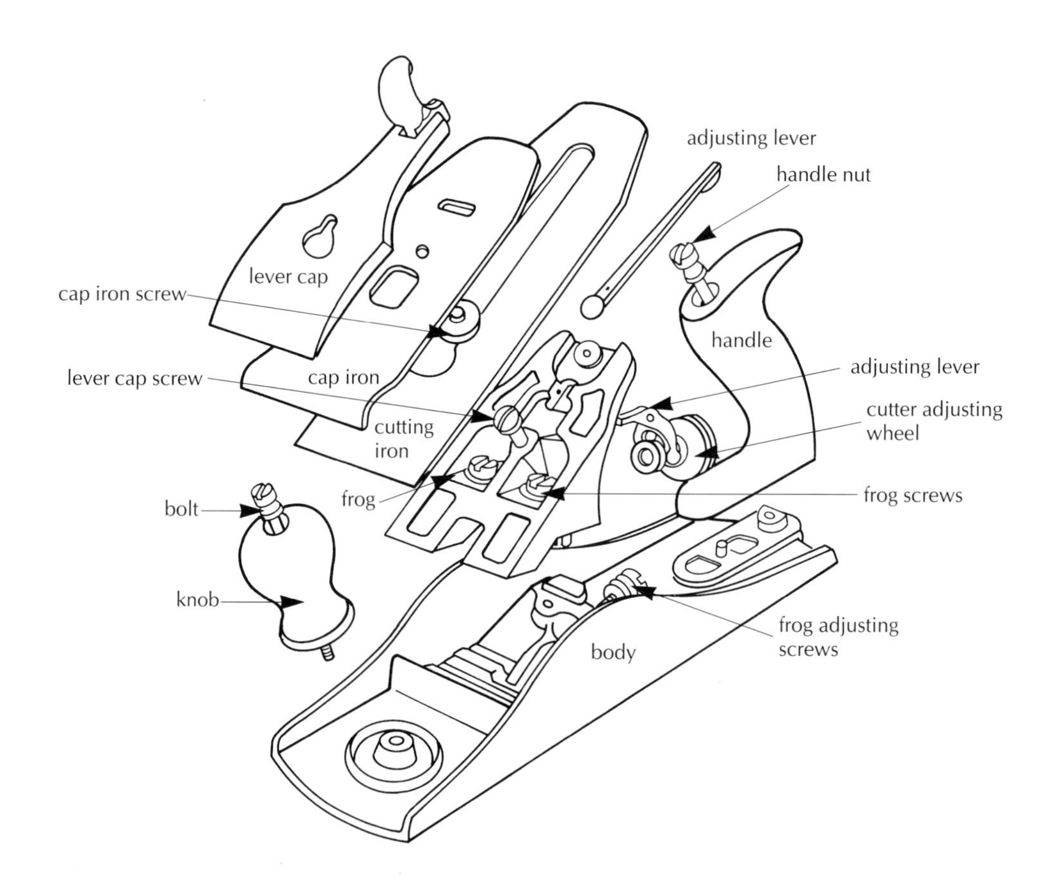

Parts of a plane.

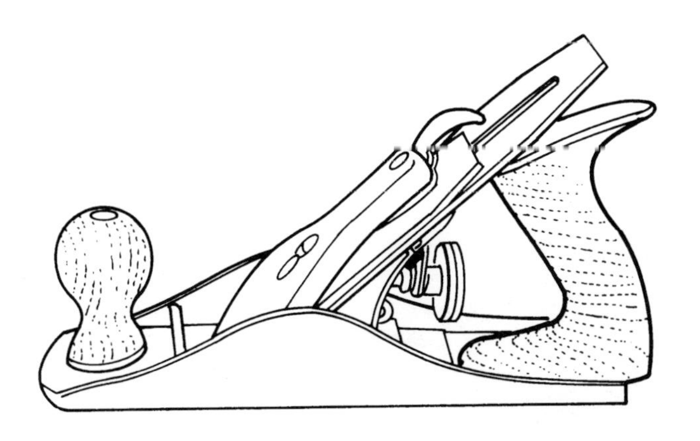

Smoothing plane.

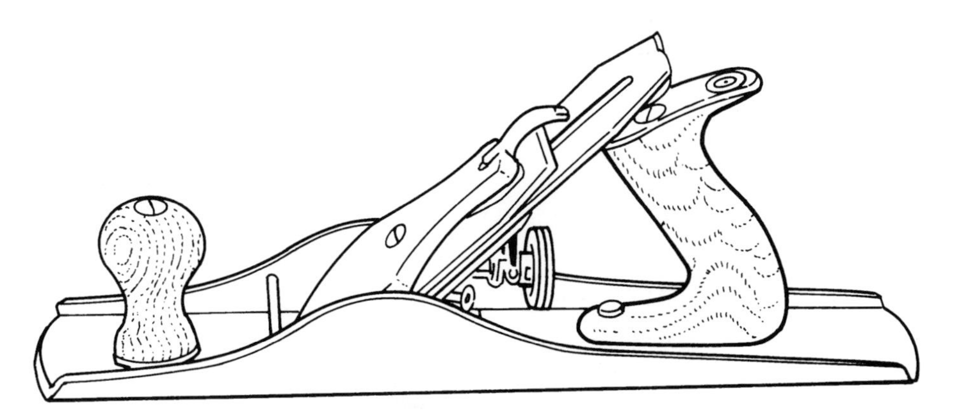

Jack plane.

finish to the piece of work. Their sizes range from 45 × 240mm (1¾ × 9½in) to 60 × 260mm (2⅜ × 10¼in).

Jack Plane

This is used for planing relatively rough timber and for rapid stock removal. In essence the starting point of many jobs, it will flatten and square timber in preparation for more refined work. Sizes vary from 50 × 355mm (2 × 14in) to 60 × 380mm (2⅜ × 15in).

Try or Jointer Plane

This is the longest of the bench planes and apart from its length resembles the standard jack plane. It is used for planing long edges ready for jointing and for truing excessively long surfaces. Its long sole allows it to bridge lengthy uneven pieces. A common size would be 60 × 560mm (2⅜ × 22in) although longer models can be obtained.

SPECIAL PLANES

The *block plane* is a worthy little plane used for end grain planing and trimming small pieces of work. It has a single forward iron where the cutter is reversed. The best block planes have lateral and throat/mouth adjustments. They have the added advantage of being employable on plastics.

The *shoulder plane* is designed specifically for very precise and accurate planing. It is always made of metal and is useful for squaring and cutting the shoulders of

TAKING CARE OF A PLANE

Looking after a plane is important if you are to get the best out of it.

- Always check that the sole is perfectly flat by using a straight edge, such as that on a square. If it is not, return the plane to the manufacturer and ask for a replacement if it is relatively new. Alternatively, take it to a professional to surface grind for you – or true it up yourself by placing it on a flat machine table or piece of glass and carefully rubbing it up and down across a piece of emery cloth held down with double sided tape.
- Keep the exterior of the plane clear by carefully running around the edge with a draw file.
- If you replace the cap iron, avoid bringing it into contact with the front edge of the blade.

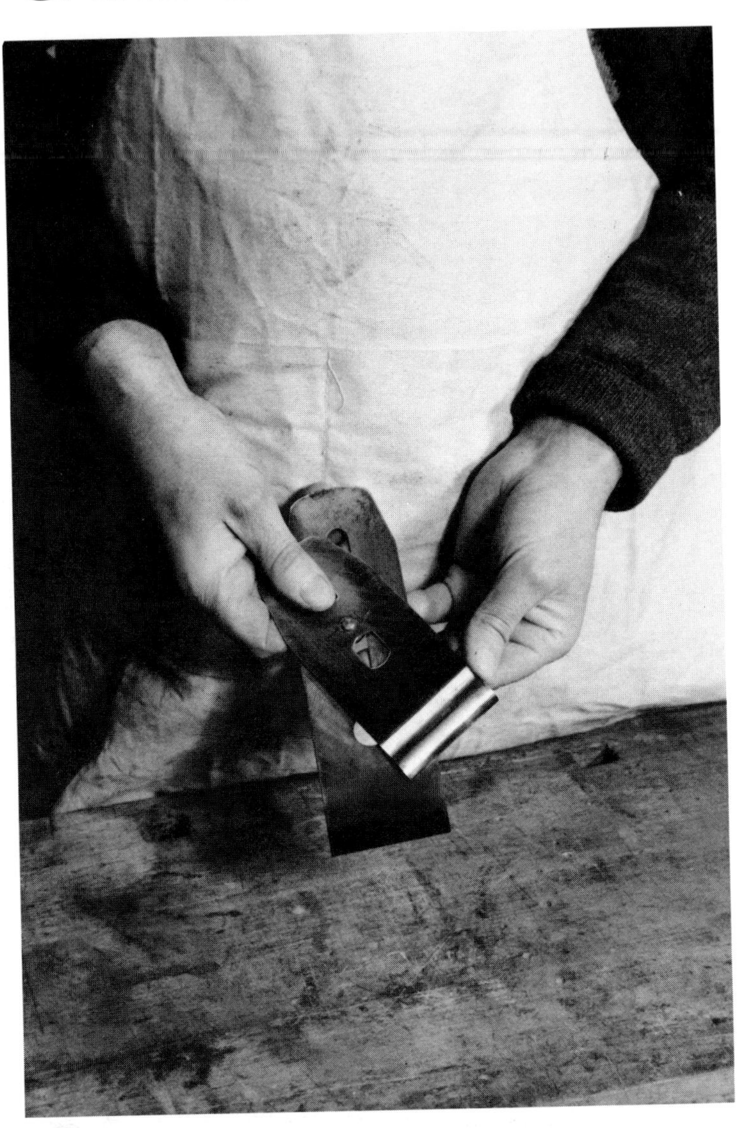

Aligning the cap iron with the blade.

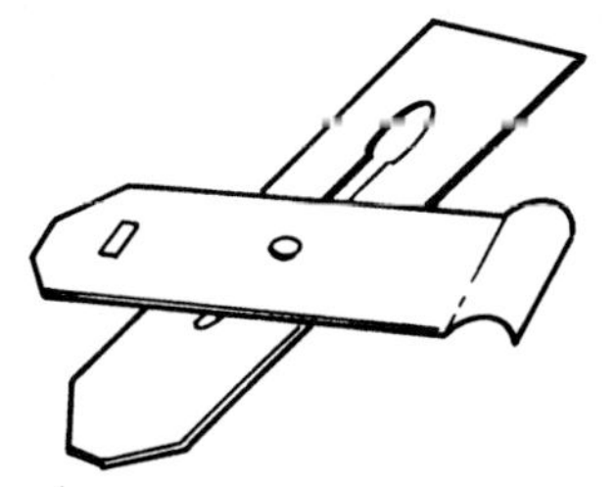

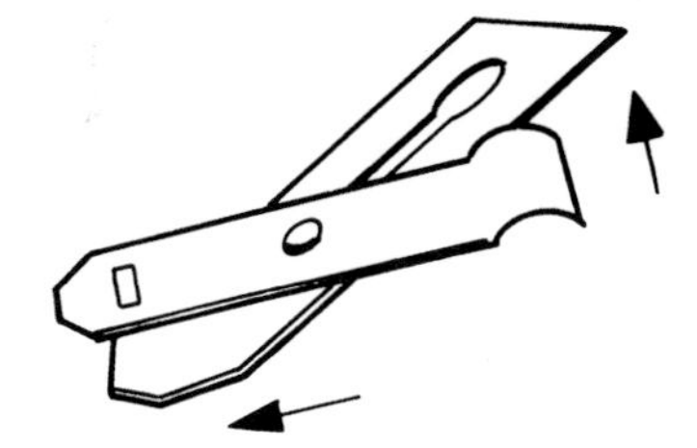

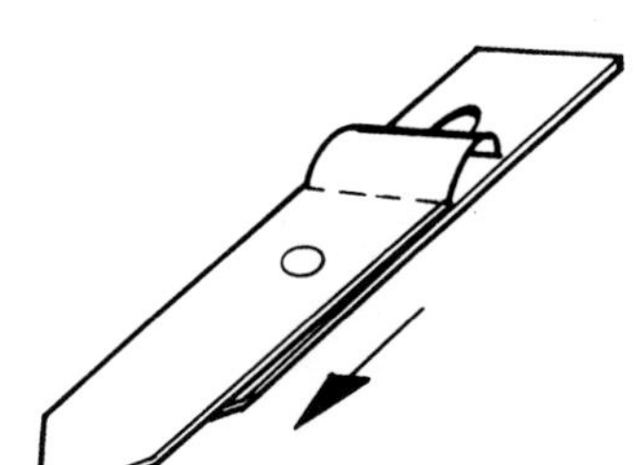

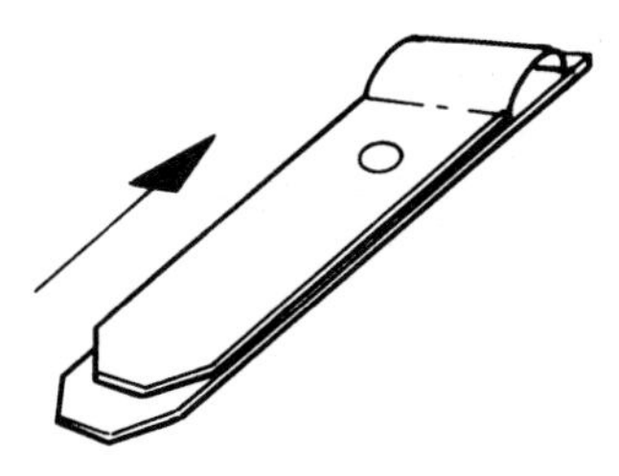

Adjusting the blade.

- When adjusting the blade laterally, hold the plane near the vertical. Screw the blade so that it protrudes through the mouth. Sight down the sole of the plane and using the lateral adjustment lever move the blade until the edge of the blade is parallel with the sole of the plane. Adjust the plane iron into the desired cut.
- When you have finished using a plane always place it on the workbench gently on its side. Banging it down may adversely adjust the cut of the blade.

Using the lateral adjustment lever to align the blade parallel to the sole. Use your apron to reflect light up the plane.

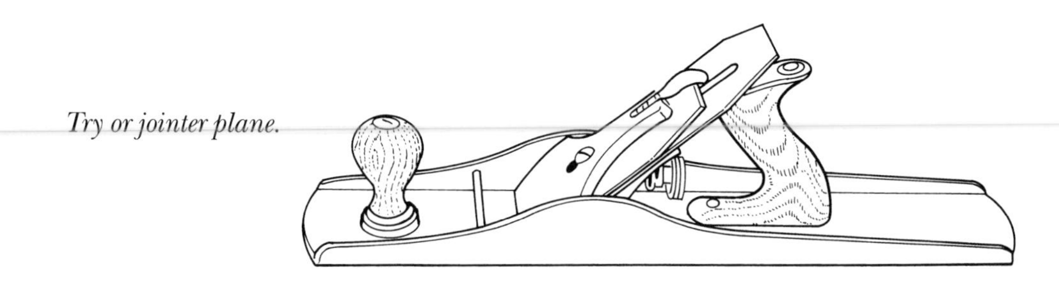

Try or jointer plane.

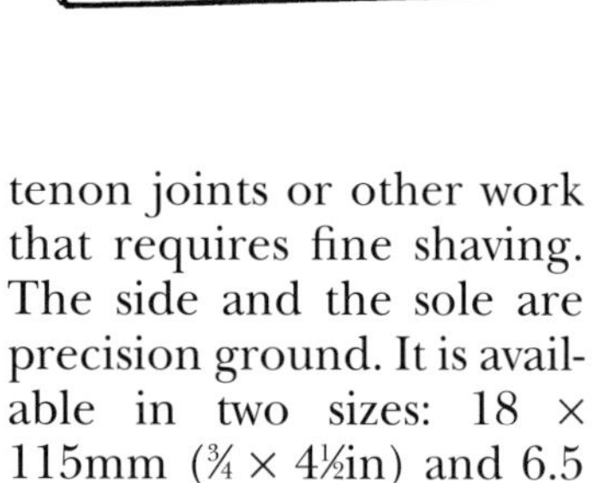

Block plane.

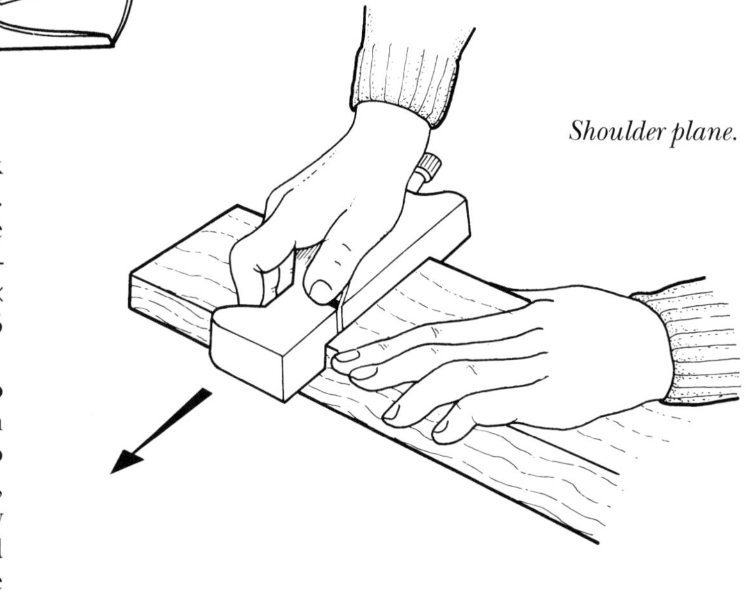

Shoulder plane.

tenon joints or other work that requires fine shaving. The side and the sole are precision ground. It is available in two sizes: 18 × 115mm (¾ × 4½in) and 6.5 × 215mm (¼ × 8½in).

The *rebate plane* is used to cut rebates up to 38mm (1½in) wide. It has two blade cutting positions, either a square or a skew cutter. It has a hardened spur for cutting across the grain and an adjustable depth stop. It measures 38 × 250mm (1½ × 9⅞in).

The *side rebate plane* is a small hand held plane used for cleaning the groove planing side of rebates. It has a removable nose for working up to the end of a stop rebate. It also has two cutters – giving it both a left- and right-hand capability – and an adjustable depth gauge. It is invaluable for

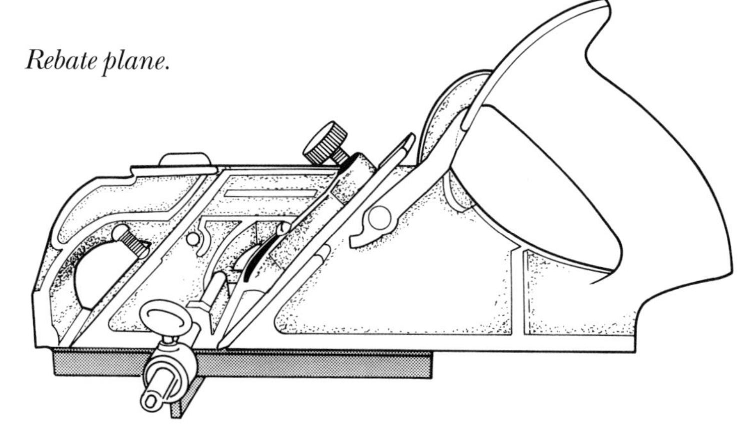

Rebate plane.

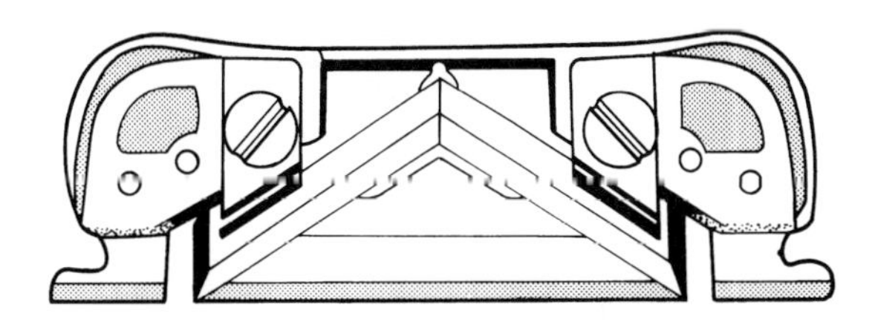

Side rebate plane.

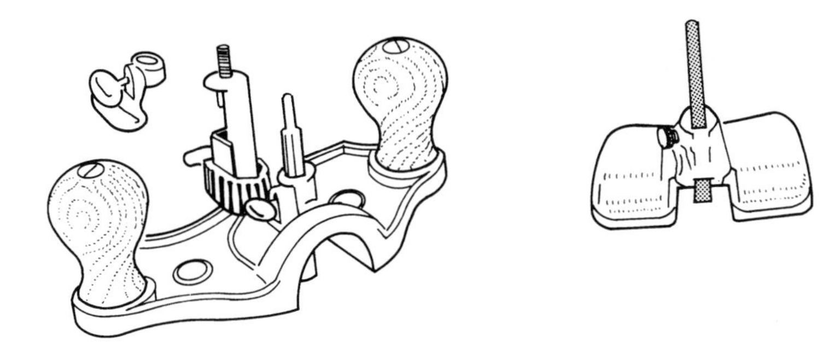

Router plane.

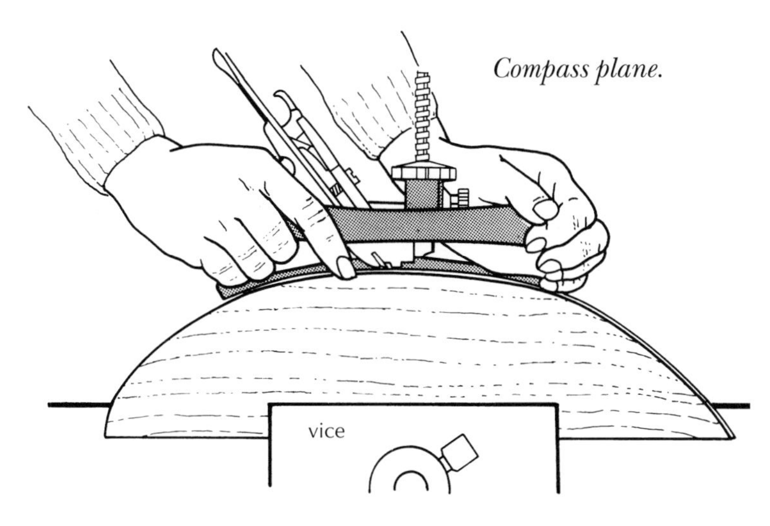

Compass plane.

increasing the width of a groove that would be inaccessible with a conventional rebate plane and it can also be used for dovetailed grooves. Make sure before it is used that the end of the cutter is just level with the edge of the plane.

The *router plane* comes in two sizes and both are used for working out and levelling grooves – such as those in the bottom of housing joints (a chisel should be used to remove excess waste first). The larger of the two router planes is more sophisticated, with the blade finely adjustable to the required depth and a precise depth gauge. The smaller plane simply has a blade adjustment mechanism. The choice depends on job requirements and the relative price differential between the two.

The circular *compass plane* is designed for cutting convex or concave surfaces. In appearance it has striking similarities to a bench plane, but the sole is made from flexible steel, which is adjusted to form a convex or concave sole through the use of a threaded bar and a sole adjustment nut. Set the plane when the sole is straight and securely lock it into place. Make sure that the cutter is always square with the work and not tilted at an angle when in use.

The *plough and combination plane* is a highly sophisticated tool that can be used for an array of purposes such as filleting, rebating and reeding. However, its primary function is for grooving – particularly making grooves parallel to the edge of a board up to

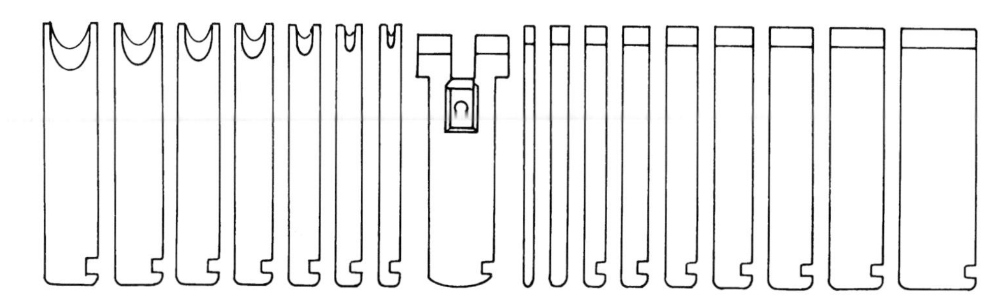

19mm (¾in) deep and up to 125mm (4⅞in) from the edge. It normally comes with a range of cutters from 3 to 13mm (⅛ to ½in). It has a cast iron frame to which the handle and the cutters are attached.

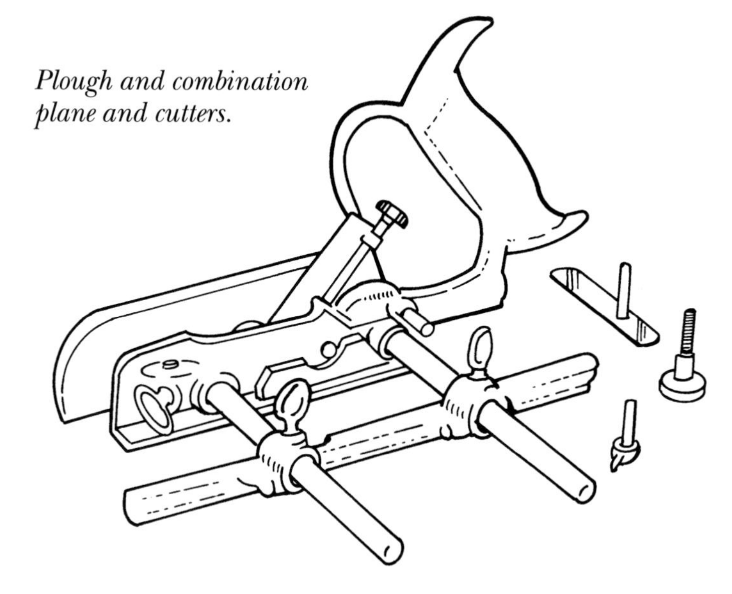

Plough and combination plane and cutters.

CHISELS

All chisels share certain common features, whatever their precise purpose, and it is helpful to start with a

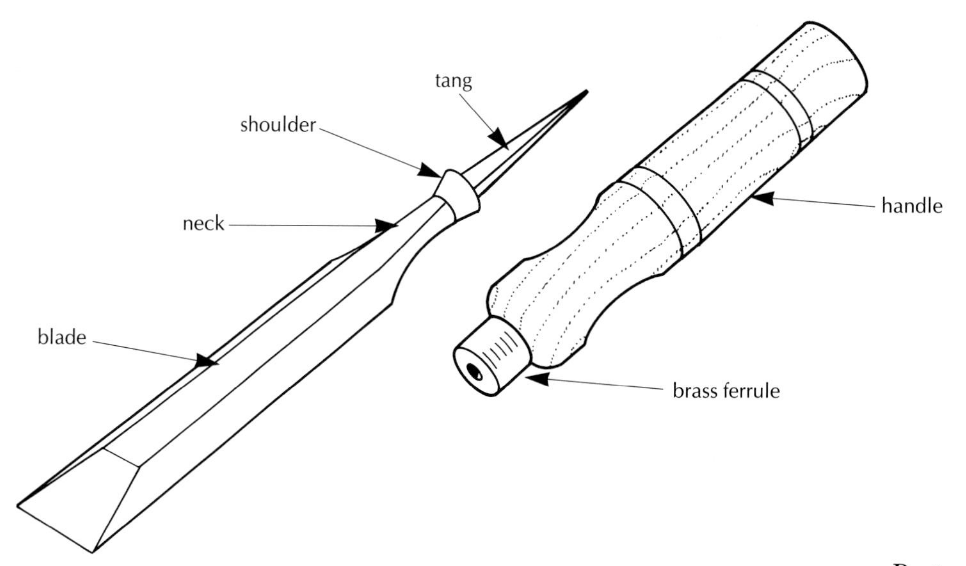

Parts of the chisel.

general overview of these characteristics. The blade of all chisels is made of hardened steel and has a carbon content of around 1 per cent. More expensive chisels can have a blade containing 14 per cent tungsten as well, to help retain the sharp edge for a longer period of time. The shoulder is forged to prevent the tang being forced into the handle when it is hit with a mallet. The tang itself is forged to a sharp point and pushed into the handle so that the blade is firmly secured.

Handles can be made in various materials – many people still prefer the feel of a traditional hardwood handle of beech, ash, rosewood or boxwood. The alternative is a high impact plastic or polypropylene. There are a vast array of chisels to choose from, some for heavy work and some designed for much lighter tasks. The shape, weight and size will depend on the purpose for which they are to be used. As with the other tools discussed so far, it is advisable to have a range of chisels that will be most commonly employed and to acquire more specialized chisels only when you need them. The chisels that should be considered essential are described below.

MORTISE CHISEL

The mortise chisel – as with a number of other traditional cabinet making tools – is gradually being replaced by powered machinery, which in many situations is quicker and more accurate than doing the job by hand. However, it still remains in use and it is important to know how it should be used correctly.

There are a wide variety of mortise chisels available, but the most commonly used ones are the *sash mortise chisel* (for very heavy work, hence its extra thick blade); the *lock mortise chisel* (the swan neck curve at the tip of the blade is useful as it permits the cleaning of waste from deep mortises); the *registered* or *standard mortise chisel* (it comes in a variety of widths and is sturdy

Using a standard mortise chisel.

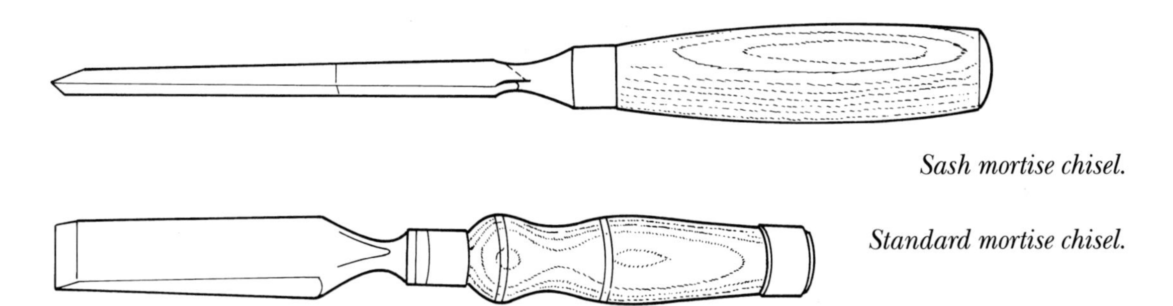

Sash mortise chisel.

Standard mortise chisel.

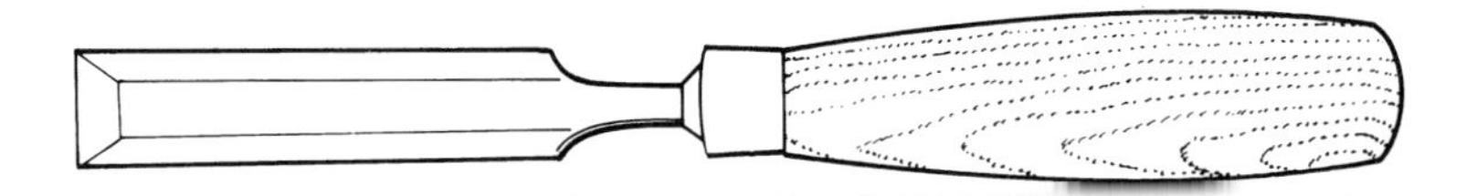

Bevel edged chisel.

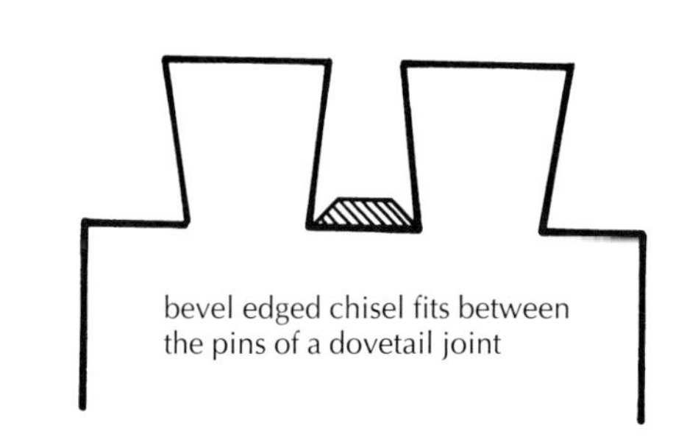

and handy for heavy work); and the *drawer lock chisel* (used for cutting hinge housings and lock mortises).

These chisels have different handles depending upon their particular applications, but all are designed to take a degree of hammering and are often struck with a mallet. When using a standard mortise chisel, always start in the centre of the mortise and work towards the end of the hole. The chisel should be held upright and knocked into the required position in small, regular steps, avoiding jamming the chisel in the joint. When the end of the hole has been reached start again by working from the centre outwards to the other end of the joint. The waste can be carefully removed from the mortise with the chisel, but be careful to avoid damaging the joint.

FIRMER CHISEL

The firmer chisel has a shock resistant leather washer for protection and comes with a hard ash handle. It is a solid chisel and the one to be used if the job requires the application of a mallet to the handle. The blade runs to about 120mm (4¾in) in length and has a range of widths from 10 to 50mm (⅜ to 2in).

The *bevel edged chisel* is similar in size and dimensions to the straightforward firmer chisel, except that it has bevelled edges. It is predominantly used for lighter work and for paring. It is not quite so robust as the standard firmer chisel, but is extremely useful for getting into awkward corners.

PARING CHISEL

The paring chisel and the *bevel edged paring chisel* come in a longer length than the firmer chisel at about 240mm (9½in). They are narrower than the firmer chisel and should under no circumstances be struck with a mallet. If any pressure is required a gentle knock on the head of the handle with a taut palm should be sufficient.

The techniques for using a firmer chisel and a paring chisel are basically the same. If chiselling a slot, the chisel should be held in two hands, the left hand in a horizontal grip at the bottom of the handle and the right hand in a vertical grip at the top of the handle. The left hand steadies the chisel, while the right hand is used to apply the pressure. It is important to make sure that the sides of the slot are sawn down accurately.

To avoid the danger of corners or edges breaking off, always chisel from both sides, turning the timber round once one side is completed. (On a heavy piece of work a firmer chisel can be used. First secure the wood onto the workbench, hold the base of the handle in the left hand with the chisel at 90° to the timber and strike gently onto the

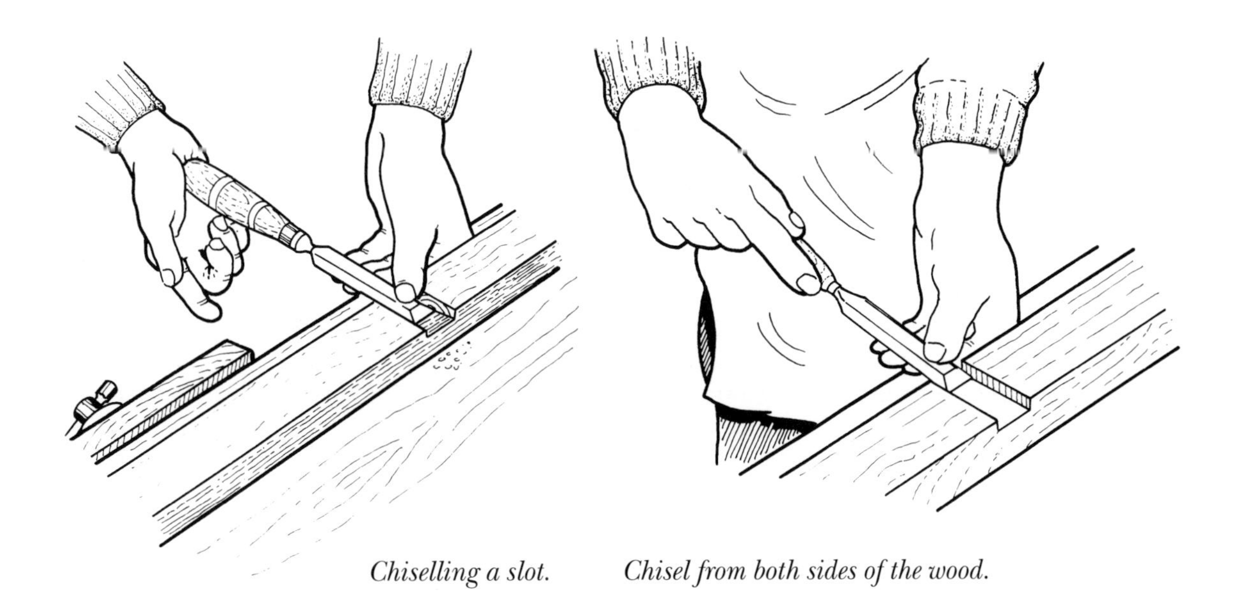

Chiselling a slot. *Chisel from both sides of the wood.*

Paring corners.

tip of the handle.) When paring corners, the blade of the paring chisel can be held between the index finger and the knuckle, which rests for support on the wood and guides the chisel into the required position. The right hand applies the pressure from the top of the handle, with the strength of the thumb doing most of the work.

DRILLS

A useful addition to the armoury would be a small selection of drills. Twist drills are made from carbon tool steel or high speed steel. They have spiral flutes to clear away waste and come in sizes from 0.4 to 13mm (1/64 to 1/2in).

Cranked paring chisel.

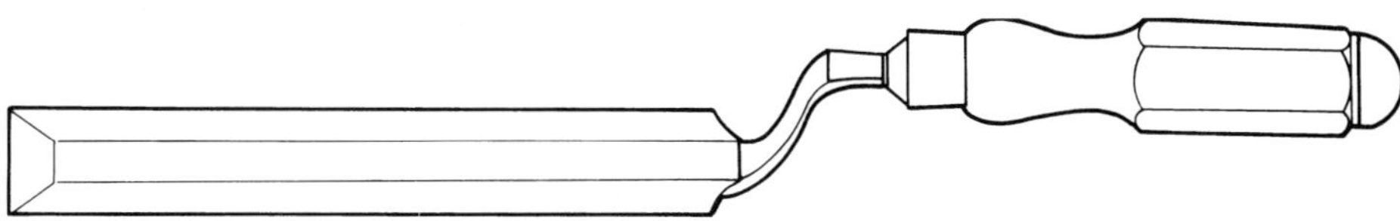

SAWS

All cabinet makers will require a collection of saws to fulfil a wide variety of purposes. There are general hand saws for preliminary work such as the cutting and slicing of timber, more specialized saws for cutting joints and saws designed to cut curves. Before describing a number of saws that would be considered essential it would be worthwhile to point out some of their general features and terminology. Sawing techniques are described in Chapter 4.

Wood consists of a mass of closely packed fibres all lying in one uniform direction – along the grain. Whenever a saw cut is made in a piece of wood, either along or across the grain, these fibres will be severed by the saw teeth. It is important that this is done cleanly and accurately, the degree of accuracy and the quality of the cut being dependent upon the nature of each individual job. Hand saws are perhaps the most common and least 'technical' group of tools, but are far from easy to use. While it is necessary to saw timber straight and square it is important to remember that if the saw is used effectively it will save enormous amounts of time on repairs and touching up.

For instance, a tenon joint sawn correctly will fit into its mortise without additional work or modification, just minimal cleaning up. By carefully cutting close to the line and avoiding the temptation to undercut, the need for finishing off with hand tools is reduced. Competence with a hand saw, as with all other aspects of saw work, develops over time, with practice and after many mistakes. However, at least some of the more glaring pitfalls can be avoided by a prudent and sensible selection of saws that are applied to the task for which they were designed.

When purchasing saws the *size* of the saw is given by the length of the blade and not the shape or size of the handle. The *tooth size* is denoted by the number of teeth per unit length – the finer the saw the more teeth it will have. The way in which the timber will be cut will also be affected by the way in which the teeth are filed and by their shape. The *tension* of the saw, which prevents the blade from being flaccid and difficult to control, is produced by specialist hammering. A saw that becomes floppy should be replaced or taken to an expert for repair.

HAND SAWS

There are three useful hand saws to bear in mind: the rip saw, the cross-cut saw and the panel saw. The *rip saw* is always used for cutting timber with the grain. When cutting keep to one side of the mark to

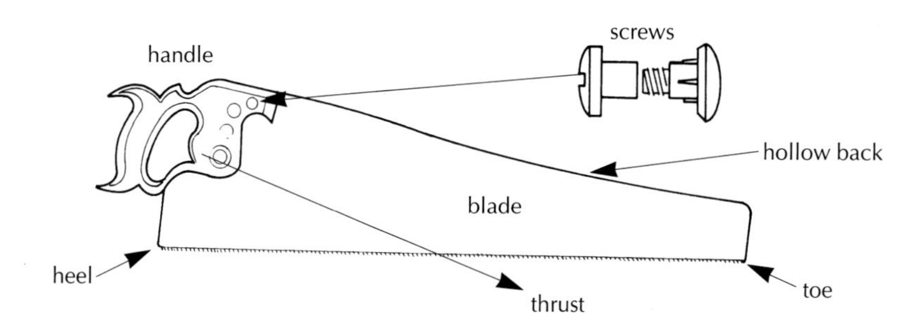

Parts of the saw.

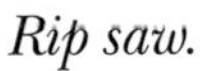
Rip saw.

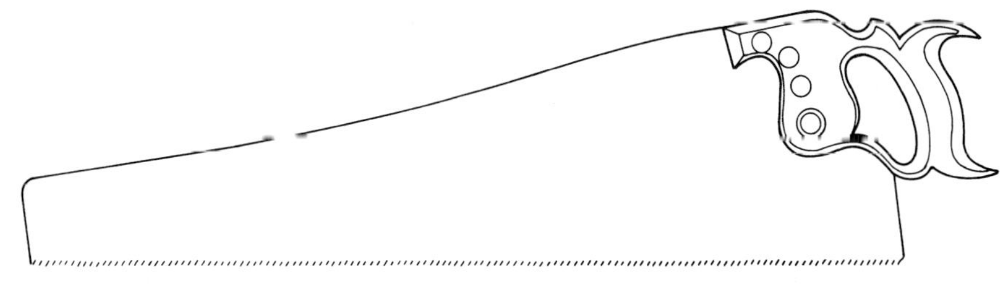

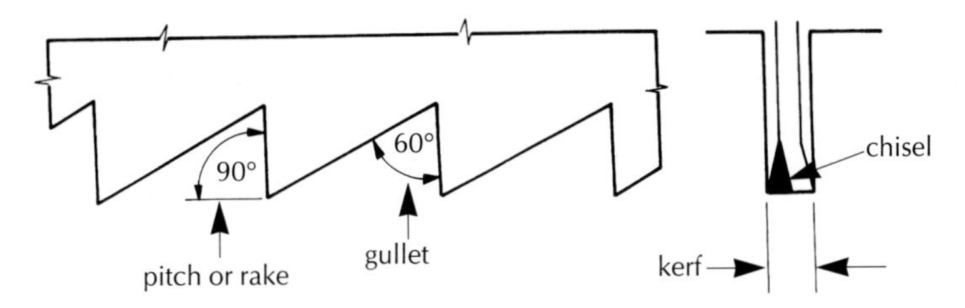

Rip saw teeth.

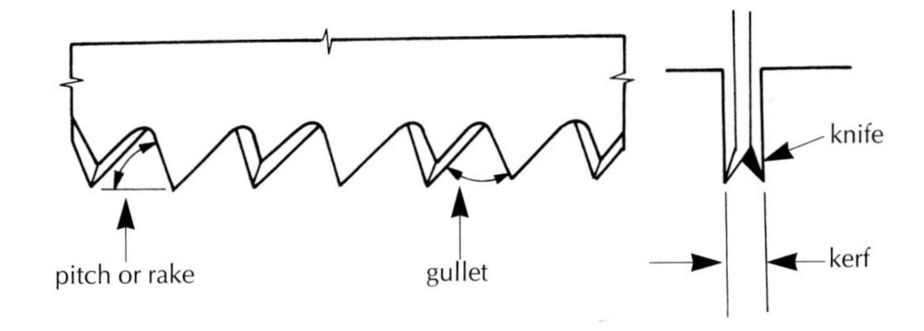

Cross-cut saw teeth.

allow for cleaning up and finishing off. These saws have chisel like points on the tips of the teeth. They come in sizes from 24 to 28in (610 to 710mm) with three to six teeth per inch.

The *cross-cut saw* is always used for cutting timber across the grain, although it can be used for cutting with the grain for which it is much slower and somewhat more cumbersome than the rip saw. With this saw the teeth are sharpened in the same way as a knife edge and therefore the action is somewhat similar to the cutting action of a knife. This gives the end product its clean cut finish. Sizes are 22 to 28in (560 to 710mm) with five to ten teeth per inch.

The *panel saw* is very important for cabinetry as it is designed for very fine cutting and is equally useful for sawing very large joints. It is essentially a cross-cut with very fine teeth. It comes in sizes from 18 to 24in (450 to 610mm) with seven to twelve teeth per inch.

BACK SAWS

Back saws are used for general work and more particularly for cutting joints. There is quite a selection to choose from and the choice is determined by fundamental requirements.

The *tenon saw* is used for an array of general purposes and for cutting larger joints. It is commonly used for cutting tenon joints, for which its relatively large, fine teeth allow a rapid yet smooth cut. It is also useful for working on plywood. Sizes are 12 to 16in (300 to 400mm) with twelve to fourteen teeth per inch.

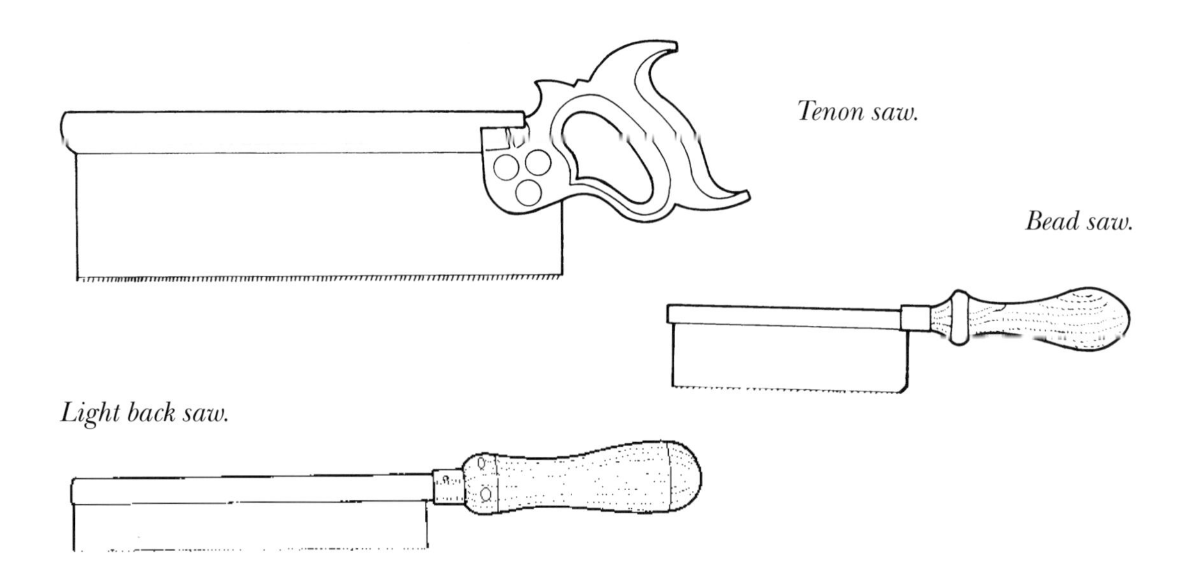

Tenon saw.

Bead saw.

Light back saw.

The *dovetail saw* is the smaller brother of the tenon saw. It is useful for fine work and cutting small joints such as dovetails and small mitres. Sizes are 8 to 10in (200 to 250mm) with sixteen to twenty-two teeth per inch.

The *bead saw* has relatively small teeth and is used for quite small work. Sizes are 4 to 8in (100 to 200mm) with sixteen teeth per inch.

The *light back saw* is the finest saw in this category and is used in very fine small scale work – very small dovetails and mitres, for example. The teeth have no set, which eliminates the need for trimming. Sizes are 4 to 8in (100 to 200mm) with twenty-four to thirty-two teeth per inch.

SAWS FOR CUTTING CURVES

There is a range to choose from, but there are features common to this group. To allow them to cut around curves the blades are fairly narrow – the width of the blade determines the angle that it can negotiate. Some of the more common saws are listed below.

The *bow saw* has two arms, across which a cord is stretched to provide the required tension on the blade, which in turn is held by two handles fixed at the base of the twin arms. It is possible to adjust the blade of

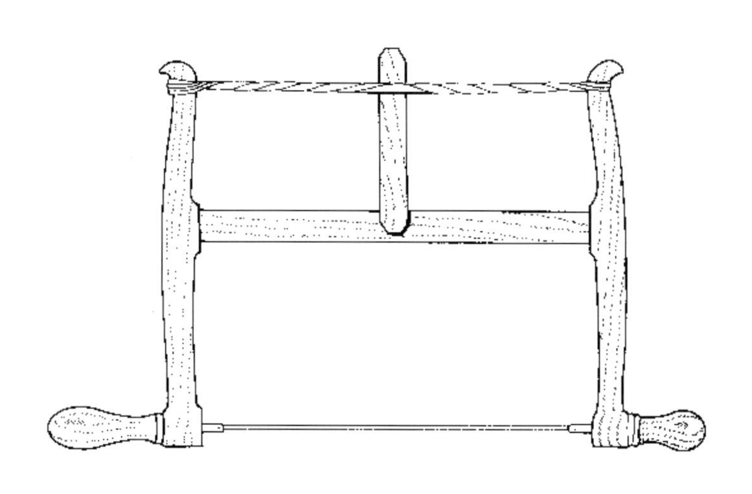

Bow saw.

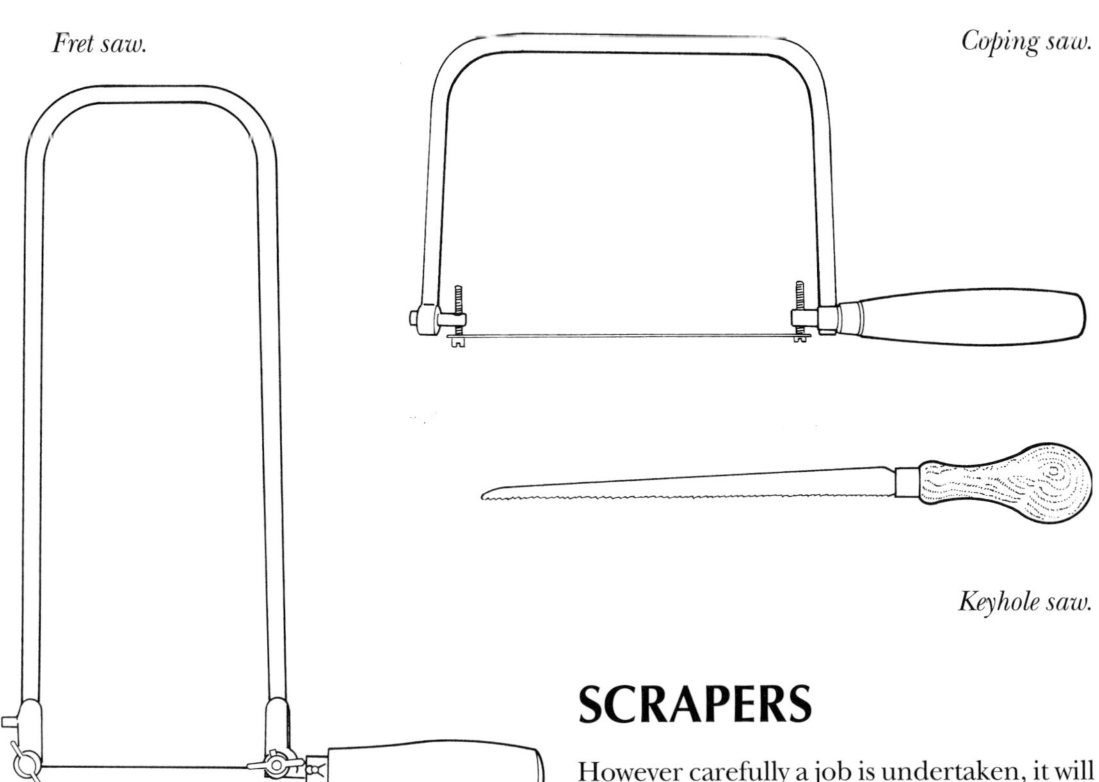

Fret saw.

Coping saw.

Keyhole saw.

this invaluable tool by turning the two handles. A range of blades from 200 to 320mm (8 to 13in) can be fitted.

The blade in the *coping saw* is tensed by the metal frame and it is handy for quick curves on shaped work of about 15mm (⅝in). The blade size is generally 150mm (6in).

The *fret saw* is a very useful tool for cutting quick curves in thin pieces of timber. The size of the frame varies from 300 to 500mm (12 to 20in).

The *keyhole saw* is a handy little saw to have around for undertaking tiny internal work. The blade comes in degrees of fineness ranging from 24 to 32 teeth per inch.

The *nest of saws* has three blades that are interchangeable and used for internal cuts and flat shapes. The blades are 250, 320 and 450mm (10, 13 and 18in).

SCRAPERS

However carefully a job is undertaken, it will need to be finished off as well as possible to give it that pristine quality we should all be aiming for. Inevitably some exposed pieces of timber will have slightly roughened edges or surfaces, or they may be clogged with dust or minute particles of shaving. Earlier plane marks may be visible or the surface could be tarnished or stained. The judicious application of a scraper will not only tidy up the job and remove any blemishes, but will leave the piece polished and glowing. In many ways the scraper will act as a final smooth planing agent.

Various types of scraper are available. The simplest and most useful merely consists of a small piece of metal and is often referred to as the *cabinet scraper*. It is simply a 100 × 80mm (3⅞ × 3⅛in) piece of carefully sharpened, tempered steel. It can be purchased commercially or home-made from a discarded hand saw blade. The scraping edge is burred (slightly raised) so that it takes off a thin shaving as it is

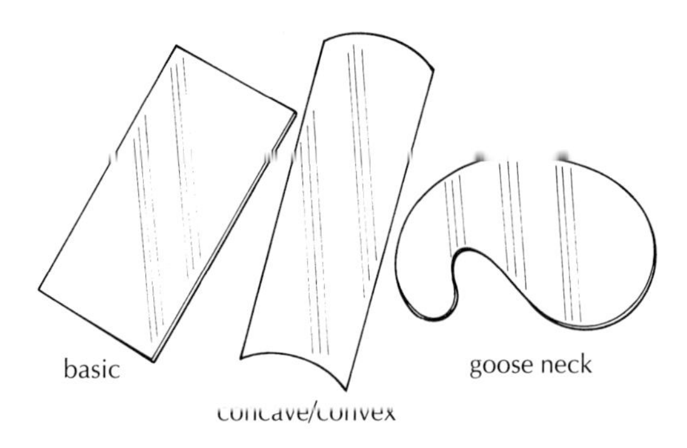

Scrapers.

TIP

It is very straightforward to use a scraper. Hold it in both hands, with the thumbs at the rear and the fingers in front. Hold the scraper at an angle of between 55 and 65° and slide it forwards across the timber in a similar action to that used when planing. It should remove a very thin shaving and leave the surface smooth and clean.

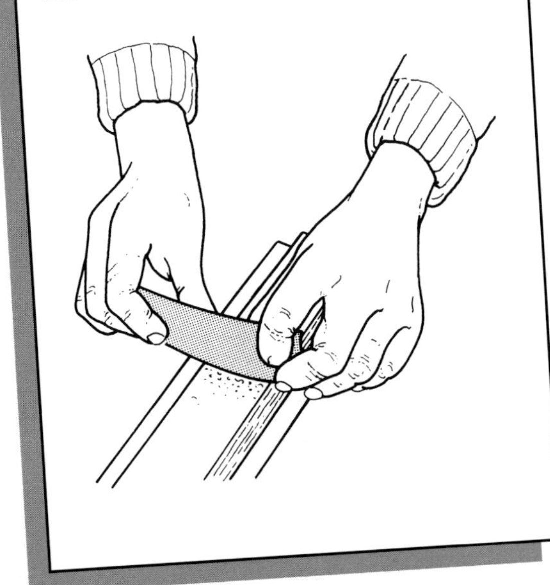

pressed along the timber. To ensure maximum effect it needs to be well prepared and cared for.

To prepare a new scraper, first secure it in the vice and use a draw file on its long edges to make it square. (The scraper is designed to follow a planing action, so it is important to make sure that its cutting edges are square.) Fine hone the edges with an oiled slipstone. Constantly turn the scraper so that a groove is not inadvertently worn into it.

Next comes the most important phase. The cutting action of the scraper is created by a raised burr and it is necessary to raise that edge. Place the scraper onto the workbench with an overhang of about 15mm (⅝in). Hold it firmly down and bring a burnisher or hard rounded gouge across the edge at a slight angle to the face. Make four or five vertical strokes. Do the same for all four edges. Finish the job by rubbing down again with a file and oilstone.

There are other types of scraper available for those occasions when the standard cabinet scraper is not deemed suitable. A *hook scraper*, with its wooden handle, is relatively easy to use. Hold it across the surface of the timber and pull it gently towards you to remove the shavings. The blades are disposable and should be replaced the moment they become blunt.

Alternatively a *scraper plane* can be employed. This tool operates on the same principle as the cabinet scraper, but is slightly easier and more comfortable to use. It simply consists of a cast iron jig into which is fitted an angled blade held in place by a metal bar and two retaining screws. The blade in this instance does not have an all round square edge, but instead the cutter is set at an angle of 45° on two edges. This curve can be adjusted using a centrally placed thumb screw.

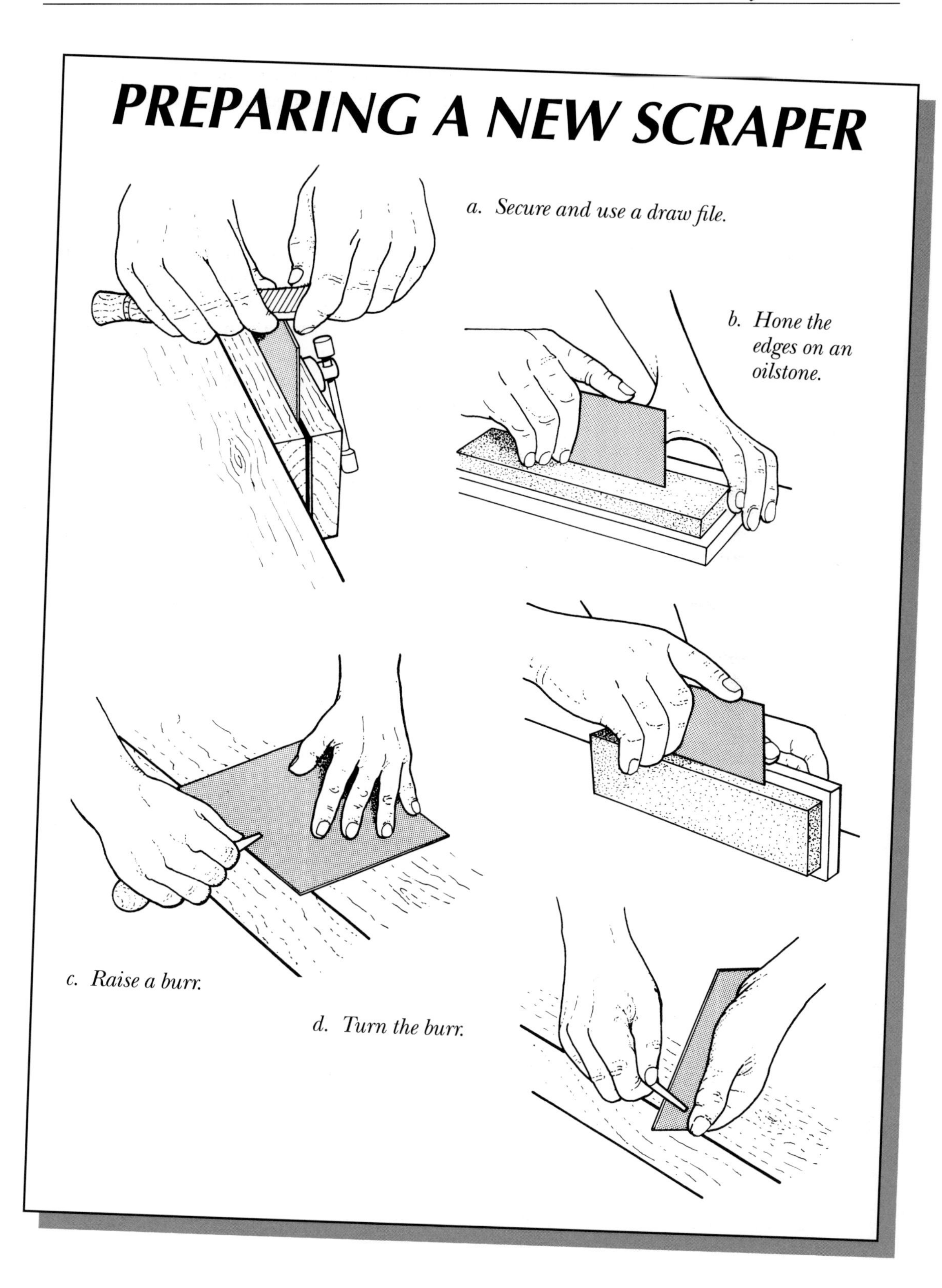
PREPARING A NEW SCRAPER
a. Secure and use a draw file.
b. Hone the edges on an oilstone.
c. Raise a burr.
d. Turn the burr.

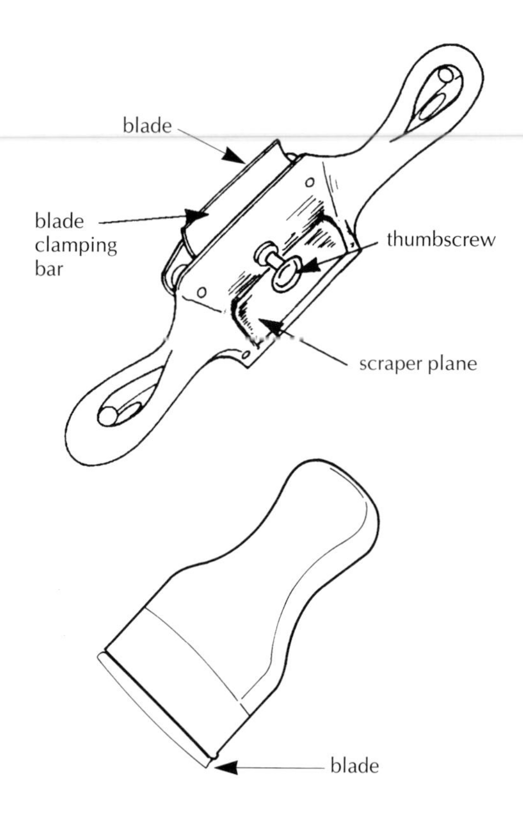

Scraper plane.

MISCELLANEOUS TOOLS

There are some very useful items that do not fall easily into any of the previous categories. The emphasis here is on the essential tools that are most valuable in saving time and energy, while not being prohibitively expensive.

DOWELLING JIG

This is an extremely versatile tool that allows for multiple holes to be drilled quickly and accurately with a minimum amount of time consuming marking out. It can be used for boring holes in any direction. The piece is attached to the jig with a thumb screw, but it is wise to use a piece of scrap timber between the screw and the work to prevent scarring. It is designed for 6.3, 8 and 9.5mm (¼, $^{5}/_{16}$ and ⅜in) dowels.

SCREWDRIVERS

The *London pattern screwdriver* is robust, but tends to be a little cumbersome to use for some work. The *cabinet pattern screwdriver* is more delicate than the London screwdriver and can be used for lighter work. In reality there is little to choose between the two and it often comes down to personal taste.

The purpose of the *spiral ratchet screwdriver* is to drive in or withdraw screws rapidly. Use one hand to apply the driving force on the handle and the other on the revolving collar to hold it in place. The upper collar can be used to vary the length of the spiral, while the sleeve is set for insertion or withdrawal.

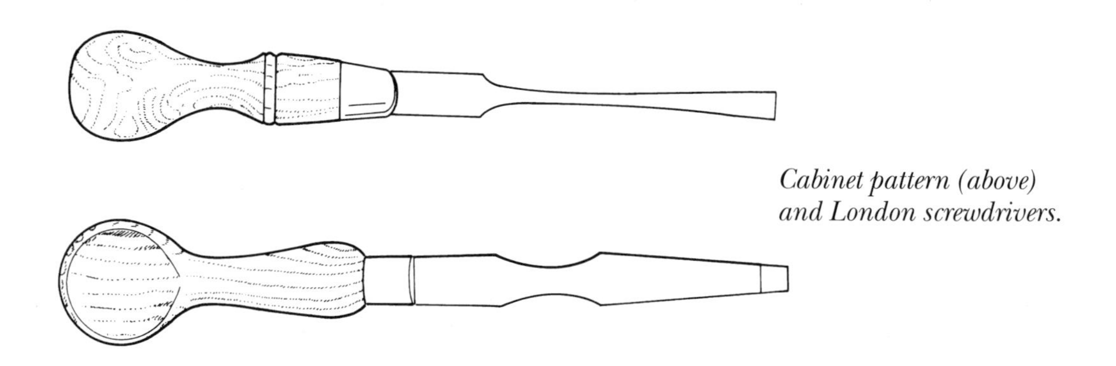

Cabinet pattern (above) and London screwdrivers.

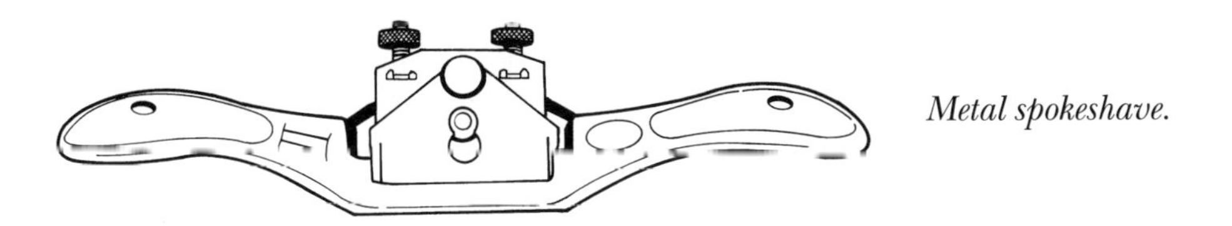

Metal spokeshave.

SPOKESHAVES

Spokeshaves can be made of wood or metal, but the preference these days is for the metal variety. The stock is generally made of cast iron and the blade of no cap iron. The blade is adjustable for the depth of cut and lateral movement. The sole or the face of the spokeshave may be either flat or round, with the flat face being used on concave curves.

SURFORM TOOLS

A variety of tools fit into this category and serve a variety of purposes: cutting, shaping, shaving, filing and rasping on a wide range of materials.

SHARPENING

To produce a quality finish to a piece of work it is essential that all cutting and edging tools are finely honed. To do this effectively it is necessary to purchase a range of stones. They are sold in a standard size of

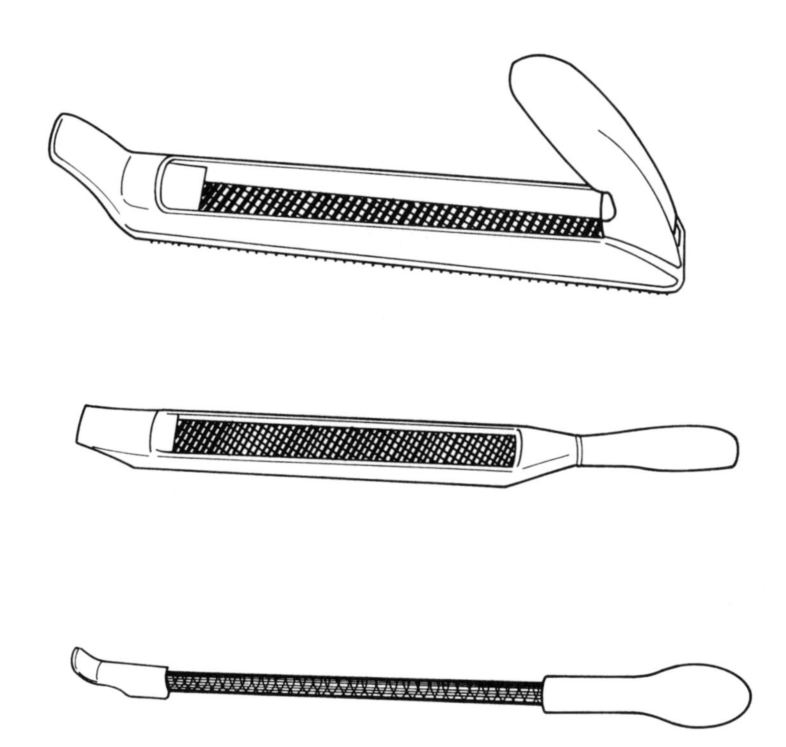

Surform tools.

203 × 50 × 25mm (8 × 2 × 1in). The quality and texture of each vary and so naturally does the price.

The most commonly used stones include *Washita,* probably the best of the natural stones available. It is of a slightly poorer quality than some other stones, but nonetheless cuts slowly to give an excellent sharpened edge and has the advantage of being reasonably priced. *Arkansas* is a superior, high-quality stone and can be expensive, though it does produce a fine cutting edge. Bear in mind that as natural stones the quality of both Arkansas and Washita can vary. *Turkey* is a very soft stone that comes in a dowdy greenish colour and is useful for very fast cutting.

The advantage of artificial stones is that they can be manufactured to a consistency that is not found in the rarer and more expensive natural stones. *India stone* is orange in colour and is produced from an oxide of aluminium. It is fast cutting and available in various grades (coarse, medium and fine). *Carborundum* is grey or black in colour and is extremely fast cutting, giving a rough finish when compared to an India. It is made by a fusion between coke and sand in an electric furnace.

Japanese water stones have become very popular in recent years and they can be inexpensively purchased from specialist suppliers. They have the advantage of not needing to be used with oil and can cut fast to produce a pristine edge. The one drawback is that they can create problems if not expertly handled as they are very soft.

Shaped *slip stones* come in natural and synthetic varieties. They are small and used for sharpening carving tools and gouges.

Grinders are used to put an angle onto chisels and planes prior to honing on the sharpening stones. They are designed to be used either dry or wet – the more favourable choice is a wet grinder because the cutting edge is cooled as it is being sharpened and it therefore does not lose its temper.

TIP

Whichever stones are purchased for sharpening cutting tools it is a good idea to maintain them in a lidded wooden box that remains closed when they are not in use. In this way they will be protected from the accumulated dust and grease of a typical workshop. If looked after they will last a considerable time.

CHAPTER

Machine Tools

There is a huge selection of machine tools available for the use of the cabinet maker. There are portable powered tools, tools fitted to the workbench, and heavy, permanently sited machine tools. As with any other piece of equipment the choice of machine will be determined by the demands of the job, cost and the available space within the workshop. A sensible purchase can save hundreds of hours of hard work and increase the quality of the finished article, but an inappropriately purchased machine can become an expensive dinosaur. So think carefully about what specific purposes any machine would be used for, how often it would be used and where it would be located. Also bear in mind the cost of electricity, maintenance and repairs before rushing to the retailer. Whatever purchase is ultimately decided upon, safety should be one of the prime considerations.

All woodworking tools are hazardous. Cut thumbs and bruised fingers are part and

MACHINE TOOL SAFETY

- Make sure you have been adequately trained.
- Make sure the working environment is safe. Check that there is sufficient room to work – cramped working conditions are dangerous. The area should be clean and clear of waste material on the table and floor to prevent slipping.
- The workshop should be warm and well illuminated.
- Long hair should be tied back. Loose clothing, which can get snagged in moving parts, should be avoided. Shirt cuffs should be buttoned.
- Always use ear protectors, safety glasses and safety boots.
- Have good dust extractors to remove harmful particles from the atmosphere and maintain good visibility.
- Check and adjust all safety guards.
- Check that tooling and cutters are correctly set and fitted.
- Never adjust the machine while it is working. Always isolate the power supply first.
- Always use a push stick.
- Use only sharp cutters and blades. Blunt tools cause accidents.
- Do not store tools, work or any piece of equipment in a position where it can fall onto the machine.
- Switch off the power to isolate the machine and wait for it to stop before leaving it.
- Keep children away from all machinery.

parcel of the craft, an occupational hazard that is not pleasant but has to be accepted. Any power tool – whether mounted on a workbench, fixed into the floor or portable – carries extra risk because of the power supply, which will continue to operate regardless of any accident that occurs. It is therefore all the more crucial that powered tools are treated with respect.

Most accidents are caused by carelessness, when the operator ceases to follow the manufacturer's guidelines regarding how the machine is to be used – perhaps the guard is not pulled down or the timber is not securely fixed. Check the electric lead as it feeds into the mains, and do not leave the wires trailing across the floor, near water or adhesives. If the workshop is in or around the house, make sure it is secured against inquisitive children. Above all make sure that the machine is regularly checked and maintained. Important safety guidelines to follow when operating any powered equipment are given on the previous page.

CIRCULAR SAW

This is an essential tool for cutting a wide range of timbers. Generally speaking it can be adjusted for depth and angle of cut. A fixed guard covering the blade means that it should be very safe to operate.

Ripping timber using the circular saw.

Use a push stick for the last 300mm.

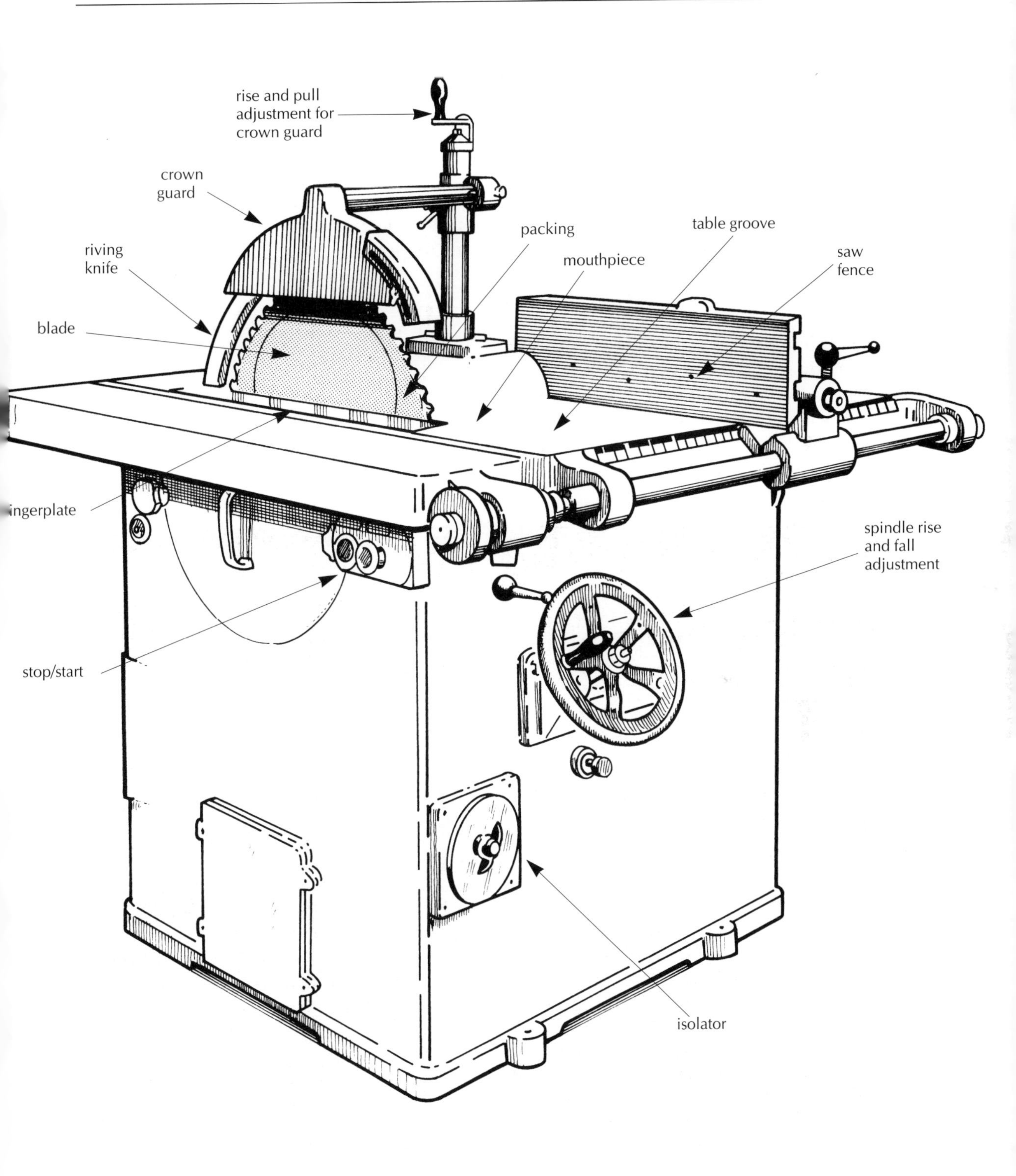

Circular saw.

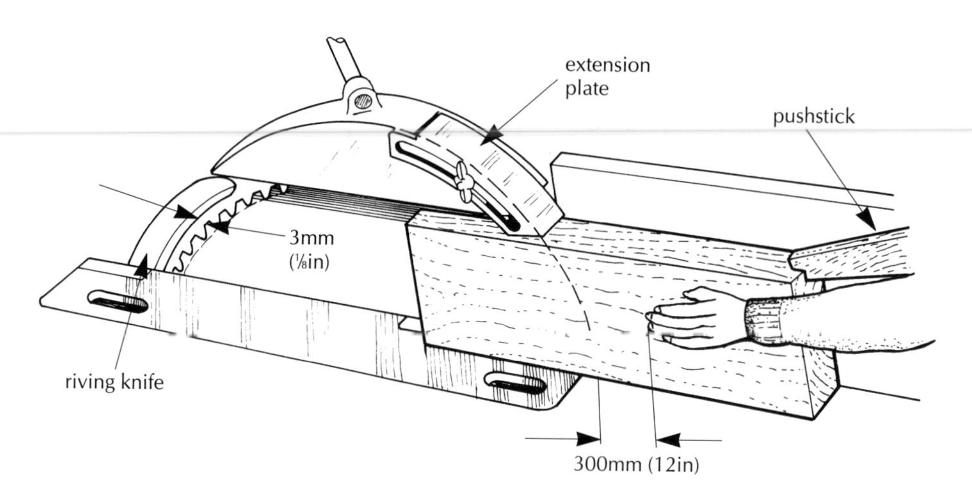

Deep cutting.

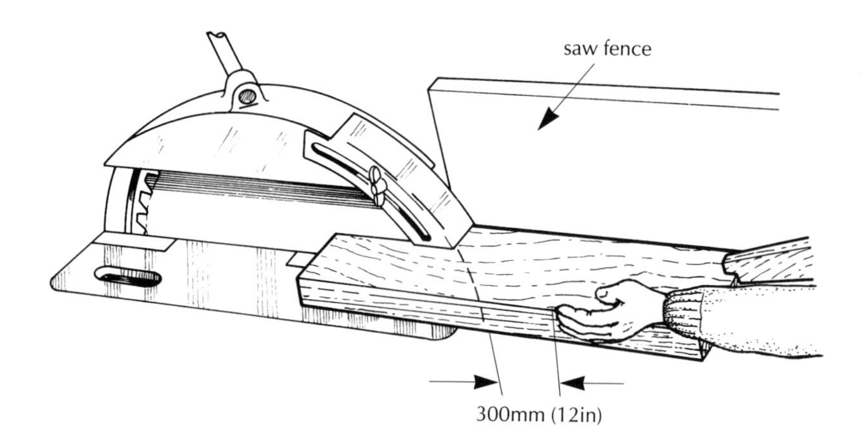

Cutting to width.

The riving knife is made from spring tempered steel and is used to prevent timber binding on the saw and also acts as a guard to the back of the saw teeth. It must be securely fixed. Normally the riving knife rises and falls in time with the saw blade.

The work table should be flat and ideally made of cast iron. Cheaper work tables are of a fabricated construction. The saw fence acts as a guide to make sure that the timber is parallel; it should be constructed so that it is rigid and sturdy.

DIMENSION SAW

This is similar in construction to the circular saw except that as the name implies it is for cutting to more precise dimensions, and it also has a sliding table. When cross-cutting use only the fence on the sliding table and when ripping make sure that the toe of the fence is in line with the gullet. Never use both fences together. The hands should be positioned so that if you slip they will miss the blade. The push stick should be made of a suitable straight-grained hardwood.

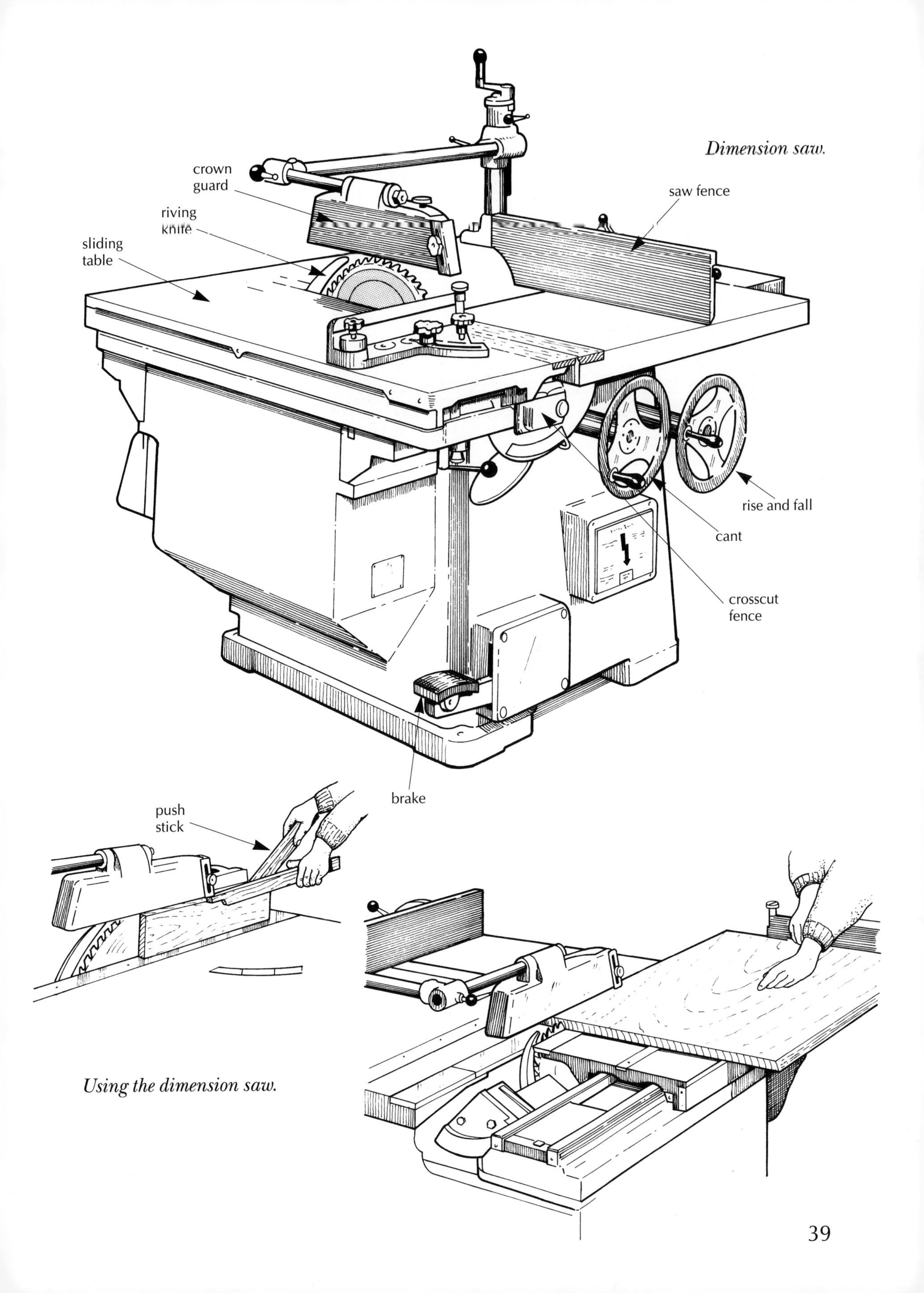

Dimension saw.

Using the dimension saw.

Squaring and cutting to length using a dimension saw.

BANDSAW

Bandsaws are most popular with many amateur cabinet makers, possibly because they are relatively safe to operate. The thrust of the blade is downwards towards the table and there is thus no fear of a kickback to the operator. It is also an eminently flexible machine and can be used for cross-cutting, ripping and cutting curved shapes. There is relatively little wastage.

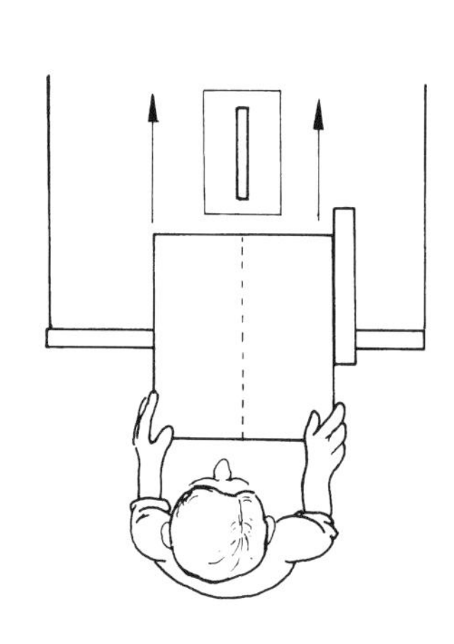

Stance for safe use of the saw.

Hands should be on the fence.

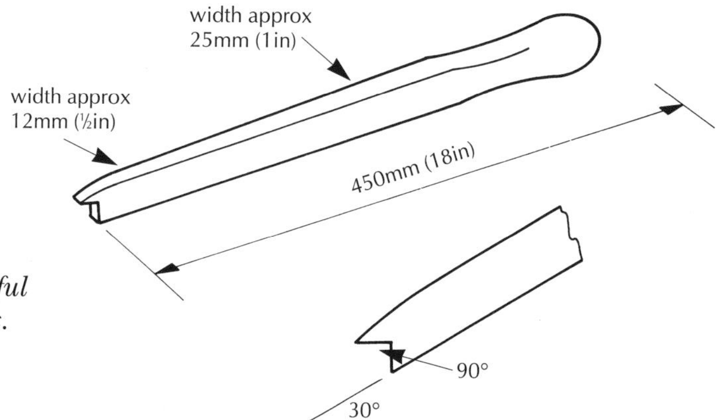

A push stick is a useful safety device for saws.

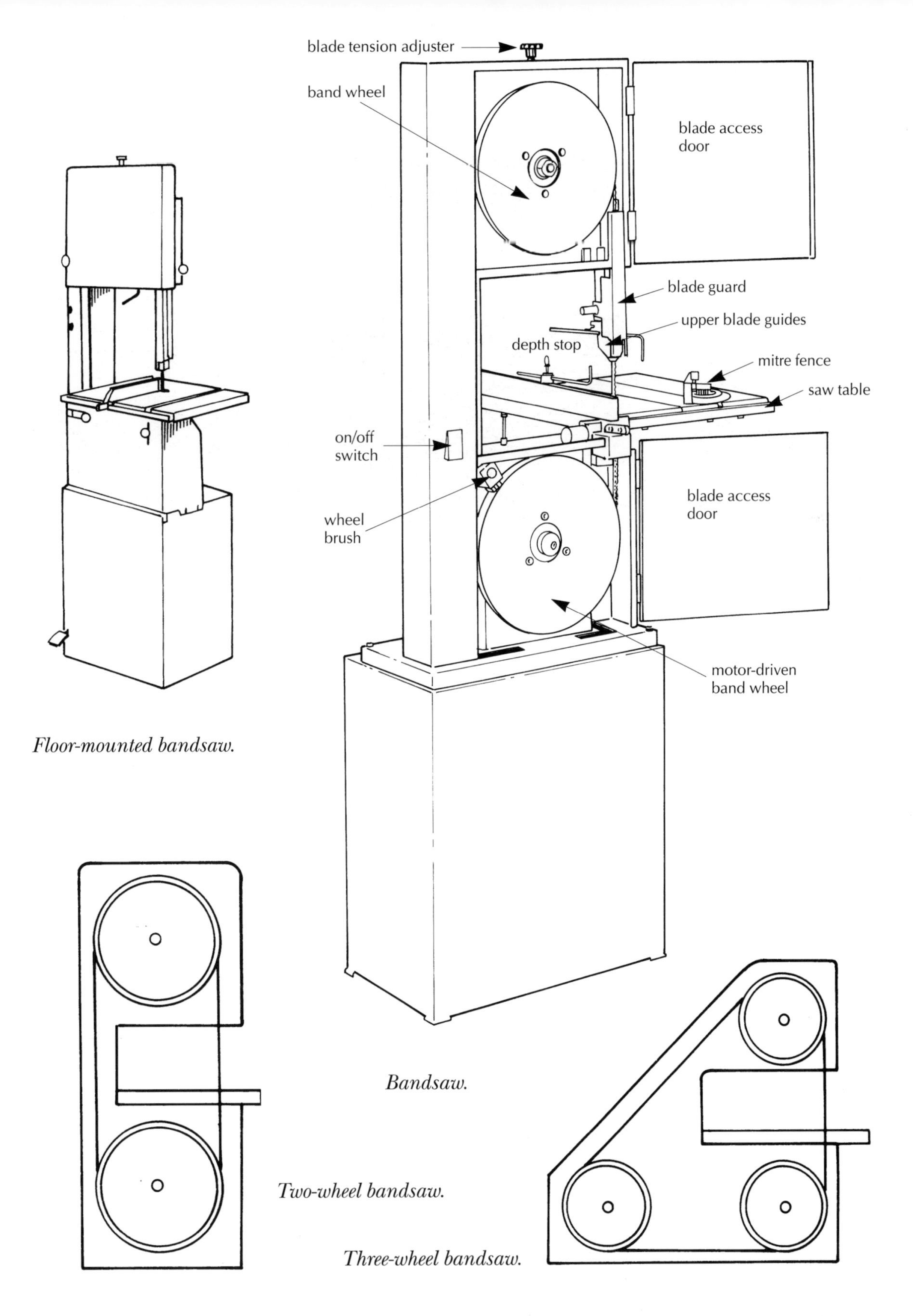

Floor-mounted bandsaw.

Bandsaw.

Two-wheel bandsaw.

Three-wheel bandsaw.

Cutting a curve using a bandsaw.

Bandsaw guides (below).

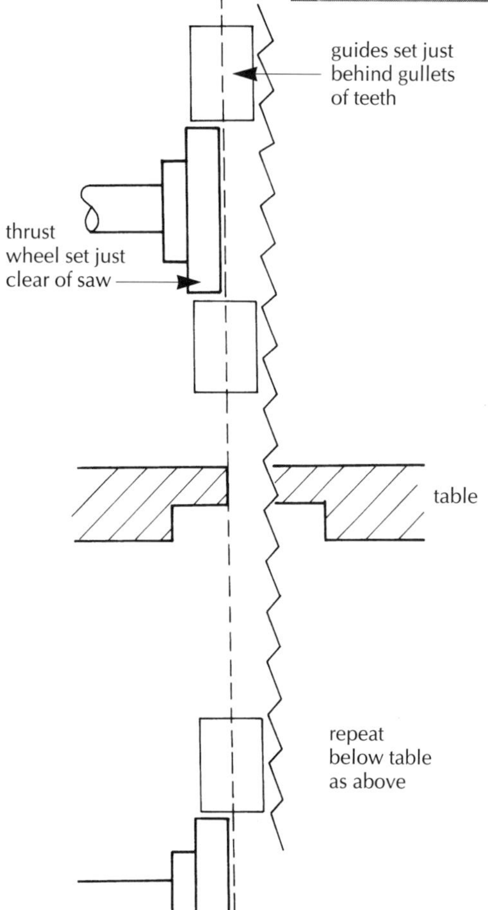

There is a 2mm (1/16in) or 3mm (1/8in) kerf. The bandsaw can have two or three wheels. Three-wheel bandsaws normally increase the throat size, although the blades on the three-wheel type tend to fatigue more quickly than on its two-wheel counterpart.

NARROW BANDSAW

This gets its name from the saw blade, which is in the form of a continuous band or belt that runs around two pulley wheels and passes through a table. Its great advantage to the cabinet maker is its supreme flexibility. Using a narrow blade it can cut a small radius, while if it is fitted with a wide rip blade it is capable of deep cutting. It is safe to cut freehand, using a template or working to a fence.

The bandsaw has guides above and below the table and with each set of guides there is either a thrust wheel or a thrust bar to lend support to the blade when cutting and to help prevent distortion across the width on narrow blades. The thrust wheel consists of a bearing attached to an

CHANGING THE BLADE ON A BANDSAW

This needs to be done regularly but is fairly straightforward:

1. Isolate the machine.
2. Open the doors that cover the pulleys.
3. Remove any existing auxiliary guards.
4. Loosen the table stiffening bar and swing it out of the way.
5. Slacken the weight and strain adjuster for the top pulley wheel.
6. Remove and fold the blade.
7. Sweep the machine clean of any dust extracts.
8. Slacken the dust wheel and guides.
9. Select and fit the appropriate blade. Adjust it to the correct tension.
10. Rotate the top pulley by hand and check to see if the blade is moving backwards or forwards. Adjust the tracking if necessary.
11. Position the guides and the thrust wheel.
12. Rotate the top wheel by hand and check that the blade is running true and is not binding anywhere.
13. Replace the guards and covers and the table support bars.
14. Close the doors.
15. Start the machine for a test run and test cut.
16. Switch off.

adjustable bar. The thrust bar is its poorer cousin and is of a cheaper construction, usually just a steel bar that comes into contact with the blade and is faced with a high carbon or ceramic disc to prevent wear. The thrust wheel should be set in line and should have just enough clearance to allow it to avoid the back of the blade when the machine is not cutting.

Take note of the safety features that apply to this particular machine. The bandsaw must have an adjustable frontal guard that fully covers the front of the blade that is exposed from the top pulley and the workpiece (except when the blade is being changed). It is also essential that in situations where the guard that covers the bottom pulley does not cover the whole of the blade below the level of the table, an extra section of guard should be fitted to compensate for this and allow no portion of the blade to be left exposed.

One other point to bear in mind is to make sure that the blade is correctly positioned on the bandsaw pulleys. This operation is called tracking, and if not undertaken will result in poorly cut timber and will ultimately damage the saw itself. It is not a job that has to be undertaken regularly, in many cases only once about every ten years. However, it is important to check the blade continuously to pre-empt costly damage. If adjustments do have to be made they can be achieved by carefully manipulating the angle of the top pulley backwards or forwards to ensure correct alignment.

SETTING UP A HOLLOW CHISEL MORTISER

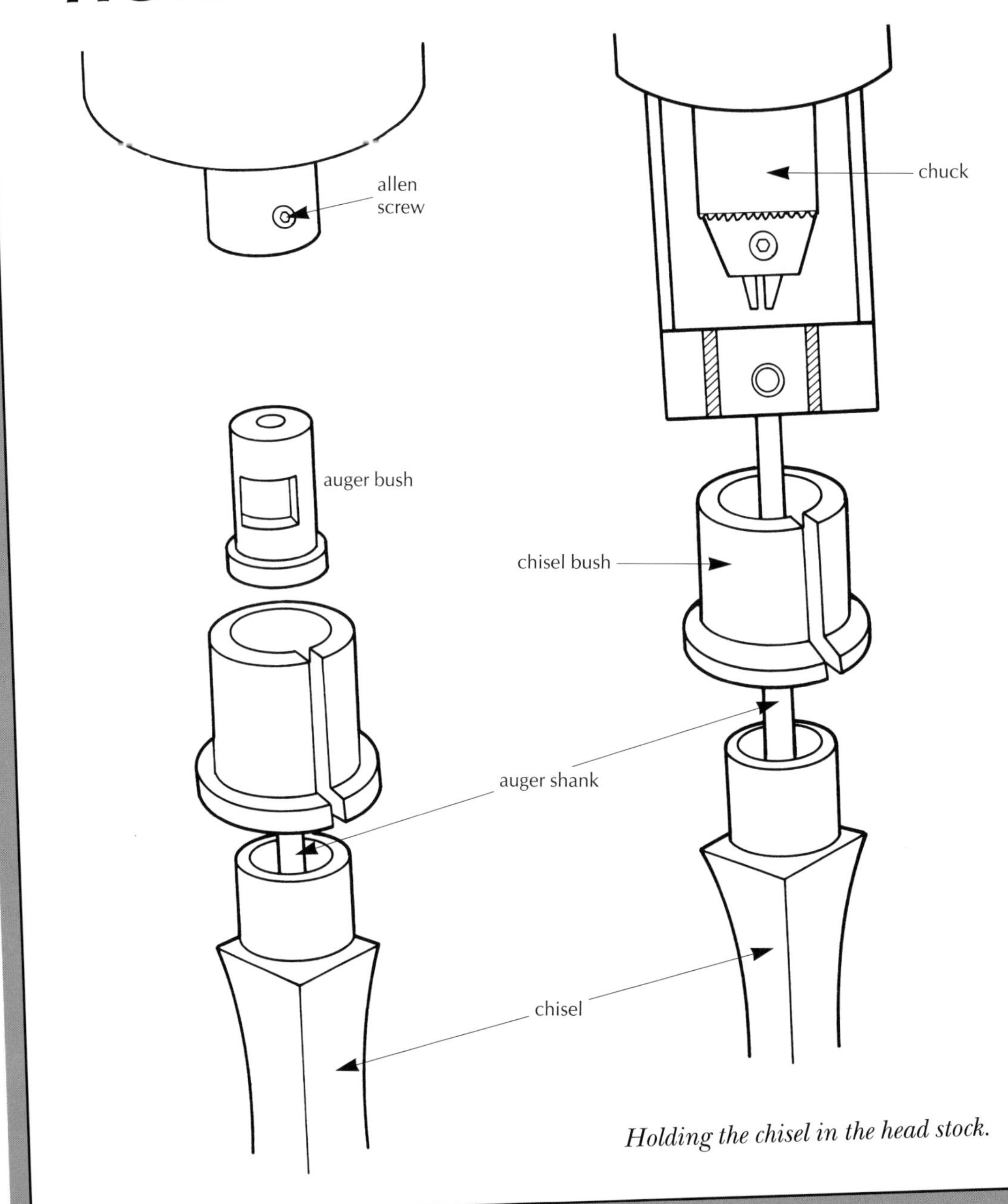

Holding the chisel in the head stock.

To set up the machine ready for use carry out the following routine:

1. Isolate the machine.
2. Select the required chisel and matching auger. Use spanners, and Allen and chuck keys.
3. Ensure that the chisel and auger are sharpened and in good condition.
4. Check that the bushes are the correct size and are in a satisfactory condition, with no burrs that could hinder the fit.
5. Place the bushes in the correct position and tighten the Allen screw to hold them in position.
6. Assemble the chisel and auger and hold them in the correct place.
7. Adjust the Allen screw to hold the auger firmly.
8. Square the chisel so that it is parallel with the fence and square.
9. Undo the Allen screw that holds the auger and then lower the auger so that its ears are just below the chisel points. Tighten the Allen screw and rotate the auger by hand; check that it rotates freely and does not rub.
10. Place the timber that is to be mortised on the table and clamp it securely in place.
11. Adjust the table to suit the mortise.
12. Set the depth stop.
13. Check the adjustment again.
14. Switch on and start mortising.

HOLLOW CHISEL MORTISER

This is an essential piece of equipment for anyone wishing to produce quality work. It enables the user to create perfect square edges and square bottomed mortises. Although it is possible to improvise by fitting a hollow chisel mortising attachment to an ordinary drill press, this is not generally advisable.

The mortiser comes in two varieties: the chain mortiser and the hollow chisel mortiser, and it is the second type with which we are primarily concerned. It cuts mortises by

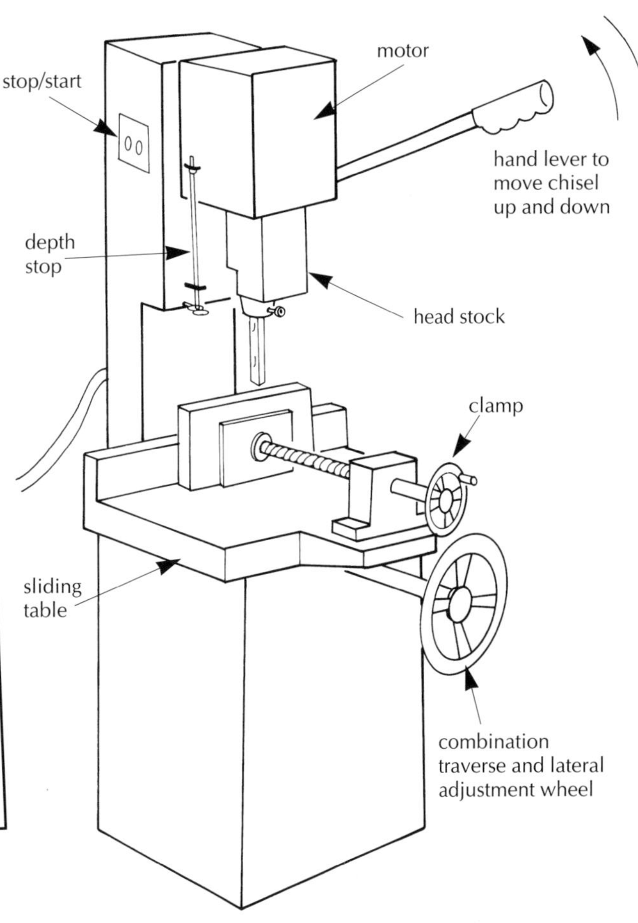

Hollow chisel mortiser.

means of a square chisel, within which an auger removes most of the waste material produced through the cutting action.

VERTICAL SPINDLE MOULDER

This is an extremely useful and serviceable machine that can undertake virtually any task required in cabinetry. It is fast, precise and incredibly versatile. If used skilfully it can mould onto curved and straight surfaces, cut dovetail housings and a wide variety of joints. However, it is also one of the most dangerous pieces of machinery in the workshop and requires careful handling and a great deal of respect.

The vertical spindle moulder is more commonly employed in moulding rebates, making grooves and in curved work, but it can cope with numerous tasks when used in conjunction with various attachments, cutters and jigs. The machine runs at a very high speed, so it is imperative that the guards are unusually robust and of the correct type for the work in hand. The jig must be of a strong, well-designed make if it is to be safe. The cutters and block must be of the correct type and set to run at the correct speed. The large variety of jobs that the spindle moulder can tackle means that it is not sufficient to have any single type of guard. Each job must be examined separately and the most appropriate guard selected.

As this is such a functional machine it is important to take care over its selection. Spindle moulders are not flimsy items. The best are extremely robust and have a strong, solid base to support the cast iron table through which the spindle travels.

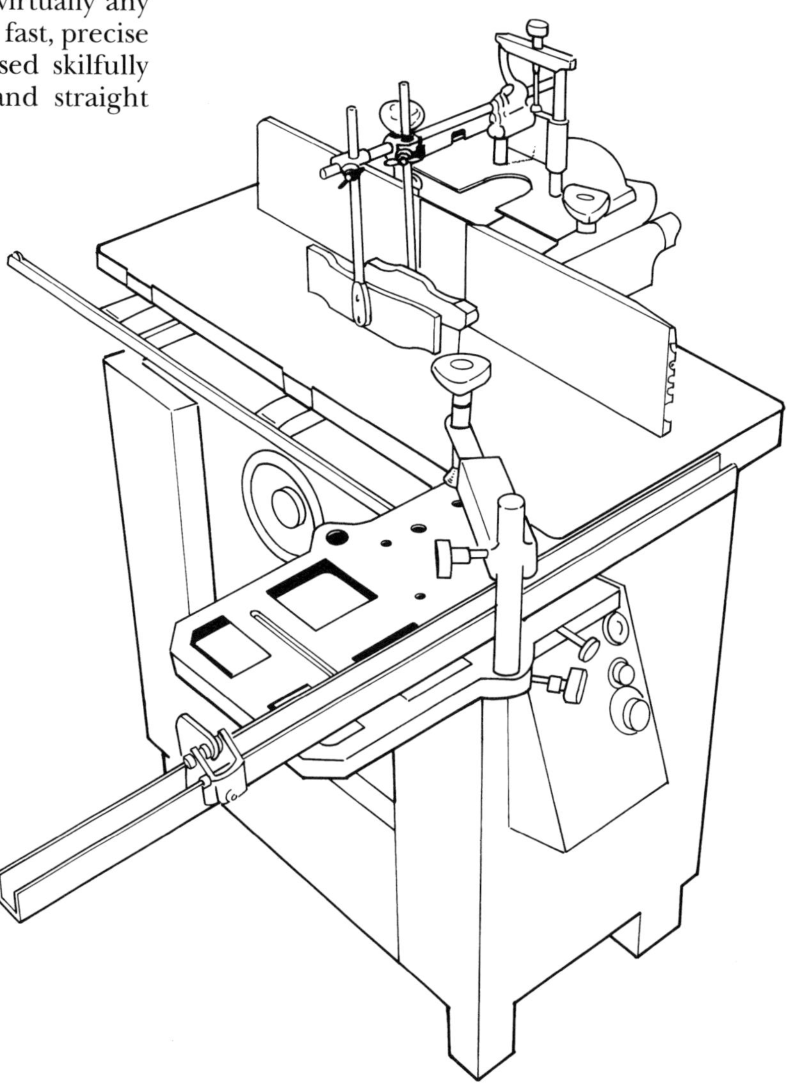

Vertical spindle moulder.

The spindle has to rise and fall through the table and on some machines it can even cut at an angle. Attachments, such as fences and guards, can be fitted to the table by means of a series of threaded holes. Other attachments available include aids for dovetailing, stair trenching, tenoning and finger jointing.

The vertical spindle moulder has two concentric rings. These make the gap between the cutting circle and the table as small as possible, thus preventing work from slipping off the table. If the tool is turned upside down it can be used as a ring fence for circular work.

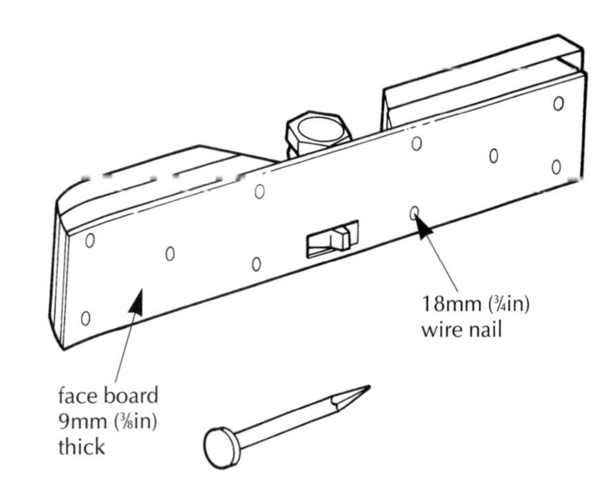

False fence.

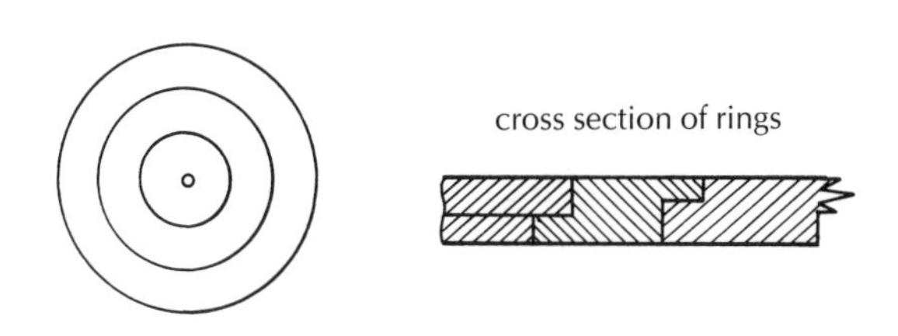

Spindle moulder rings.

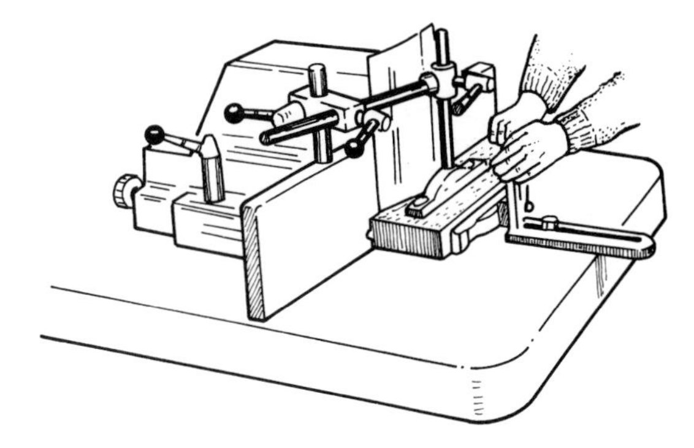
Straight fence and guard.

Some problems may be created by ill fitting rings and they should be regularly checked. There are a number of causes of rings not fitting correctly, the most common being a build up of resins or shavings or the rings having dropped and become burred. To remedy this, file off the burr and clean away the debris.

There are a number of fences that can be used. There are various straight fences, but the horseshoe type is the standard fence normally supplied with the machine. The fence incorporates a guard and extraction hood holes for the positioning of guards and fixing of pressures.

When adjustments are necessary it is possible to move the unit as a whole, while for more precise, fine adjustment each side can be moved independently. The fence plate can be adjusted laterally to

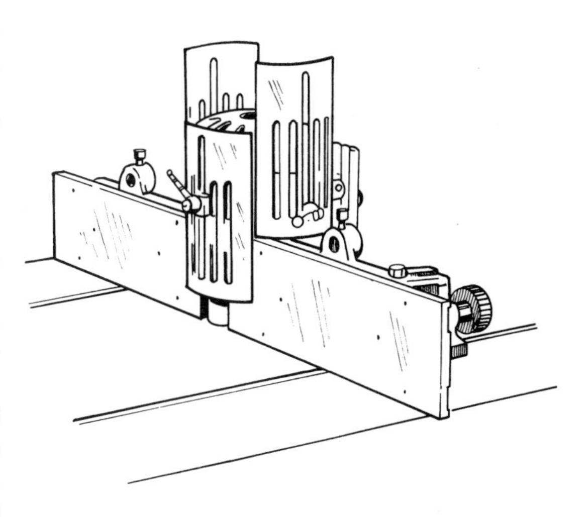
Using the straight fence.

The ring fence.

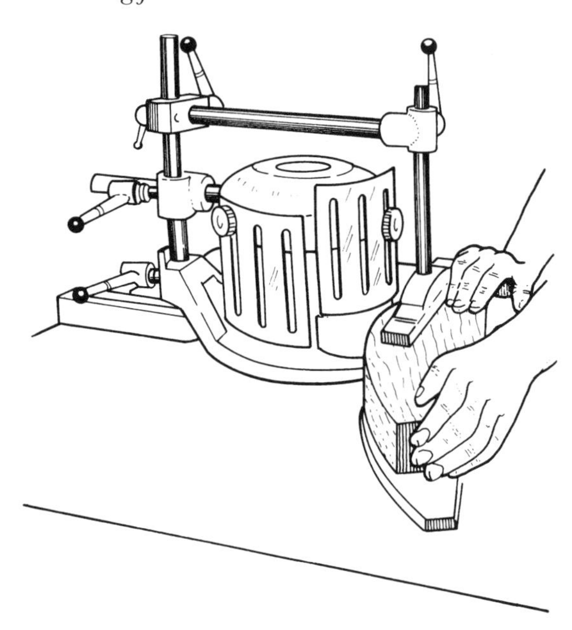

Bonnet and shaw guard plus the ring fence in use.

create a tight gap for the cutter block. For grooves and rebates a false fence can be applied, usually with wire nails. This fence only permits the part of the cutter that is providing the cutting edge to be exposed and prevents a short piece of timber from dipping between the fences.

PLANER AND THICKNESSER

It is possible to buy a separate planer and a thicknesser to undertake specific functions. Some would even argue that it is necessary to have two planers, one to true long edges and one surface and one to allow stock to be reduced to a consistent thickness. However, for many people the cost of this is prohibitive and they may not have room for both items.

An adequate compromise is to go for a combination surface planer and thicknesser. The surfacer is used first for planing up the face side before moving on to

Machine planing a surface.

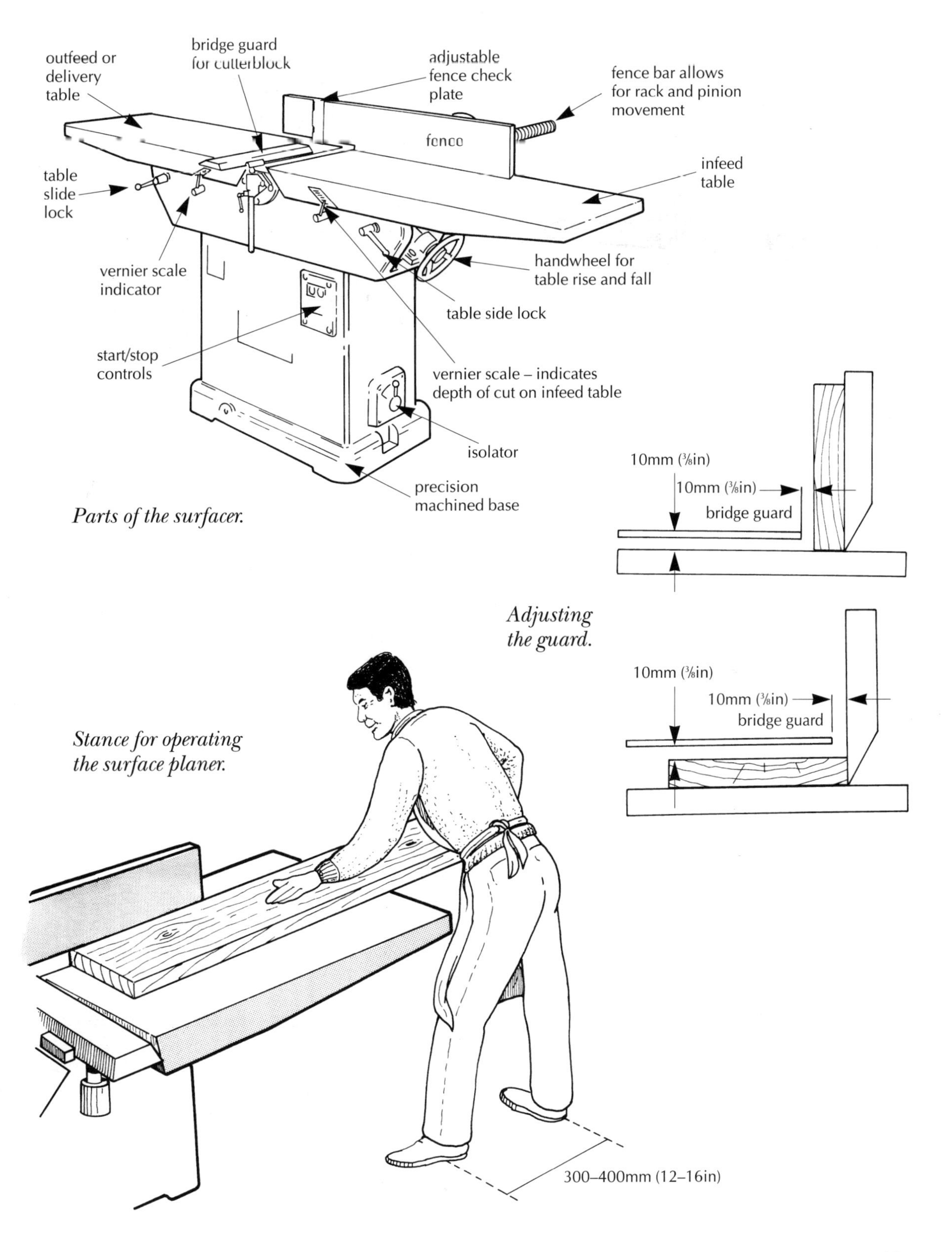

Parts of the surfacer.

Adjusting the guard.

Stance for operating the surface planer.

Creating a square edge on the surface planer.

Angle adjustment of the fence.

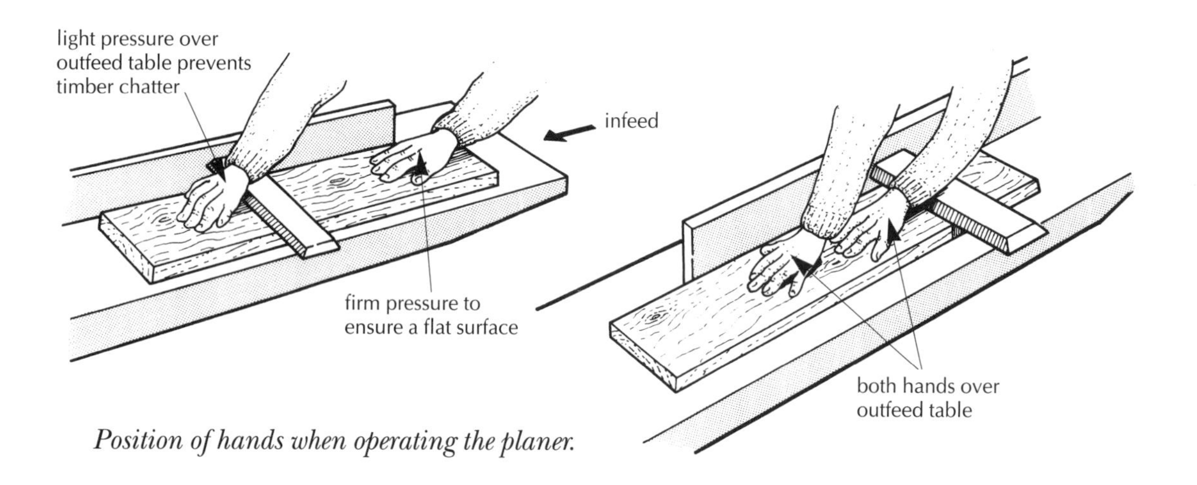

Position of hands when operating the planer.

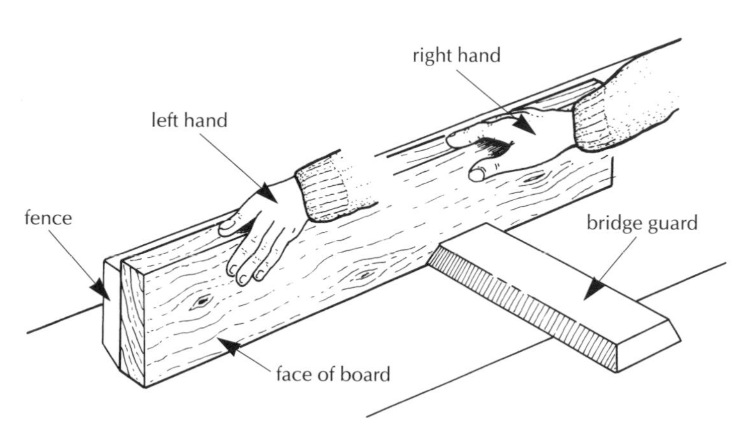

the face edge. The timber is then carefully passed through the thicknesser to make sure that it is parallel. As with all powered tools it is essential that the operating and safety rules are followed if nasty accidents are to be avoided.

Edge planing.

CHAPTER three

Materials

The range of materials available is almost endless. Much will depend on individual preferences and the budget available.

TIMBER

The most immediate and obvious material that needs to be considered is timber. This has an environmental as well as a practical aspect, for we should all be aware of the dangers facing the areas from which much of the timber is forested. The description of the timbers given below is not meant to be exhaustive, but it does cover those timbers that will be used in the average workshop.

EUROPEAN BEECH

This timber comes from the UK, parts of continental Europe and some areas of western Asia. It is white in colour or a light brown, sometimes tinged with pink, with no distinctive colour change between sapwood and heartwood. It is a strong timber and has a good resistance to splitting. It gives a good finish in hand work and has a close texture that allows it to turn well. It also stains and polishes well. It has wide applications in many areas of furniture making, but it is susceptible to attack by the pinhole borer and the furniture borer.

DOUGLAS FIR

This native of North America has also been extensively planted in the UK and in New Zealand. The wood is generally straight, but can occasionally produce a spiral or wavy grain. It has similar properties to pitch pine, with the stronger wood coming from the Pacific coast rather than the mountain areas.

In comparison to Baltic redwood, Douglas fir is 20 per cent heavier, 40 per cent harder and 60 per cent stiffer. It works readily with hand and machine tools. (Try to avoid UK-grown material, which is grown quickly and is prone to loose and hard knots, making sawing difficult.) It takes stain well and gives good results with other finishes. It is susceptible to dry-wood termites, as well as the pinhole borer and the longhorn beetle. The most important point relating to this timber is that it is the most common timber from which plywood and veneers are made.

EUROPEAN OAK

This originates in the temperate hardwood forests of Austria, France, Hungary, Germany and the former Czechoslovakia. It is yellow in colour and has a distinctive white sapwood. Its most important feature is that it is very strong in comparison to other hardwoods. When working with this timber note that it needs pre-drilling for nails and screws to avoid the danger of splitting.

Do not use paint with this wood, but it will take other finishes. It is straight grained and will plane and mould to a fine finish. The heartwood may be attacked by the pin-hole borer or the death watch beetle. It is extensively used in many areas of furniture making and as plywood and decorative veneering.

SCOTS PINE

Also known as the Norway fir, this tree is widely distributed across much of Europe. This widespread distribution leads to the timber varying in its qualities depending upon where it is grown. This is particularly pertinent to texture, density and knotting. When seasoned, the heartwood is a pale reddish brown colour, while the sapwood is lighter.

Scots pine is a fairly strong timber and it generally works easily and cleanly in most circumstances. It takes nails and screws well and can be effectively stained, also giving good results with paint, varnish and polish. It is often used for the manufacture of plywood.

AMERICAN MAHOGANY

This comes from Brazil, Peru, the Caribbean and Central America. It has a light red-brown colour that darkens in moderate light, but fades in strong light. It is straight grained and works very well with both hand tools and machine tools; it also planes to a smooth finish. It receives nails, screws and glue well and accepts all finishes without difficulty. It can be attacked by the furniture beetle and powder post beetle. It is extensively used in all aspects of furniture and cabinet work, but the only drawback with this timber is its high cost.

TEAK

Teak comes from India, Myanmar, Indonesia and Thailand. It is yellow-brown in colour and often comes with dark streaks that darken further when exposed to light. The sapwood is a distinctive white or pale yellow. It is a very long-lasting timber that works well with hand and machine tools, but dulls their cutting edges. Gluing is difficult due to its greasy nature, although it does polish reasonably well. Teak is widely used in furniture making, especially for interior joinery and fittings and is useful for garden furniture because of its robust nature. However, as with mahogany it is expensive to buy.

SYNTHETIC MATERIALS

Synthetic materials fall into three categories: fibre boards (such as hardboard), particle boards (such as chipboard) and laminated boards.

HARDBOARD

The majority of hardboard is produced by a wet process. When suspended in water, the fibres lie in such a way as to create a felt mat when the water is drained away. The remaining moisture is removed by putting the mat through presses. Hardboard can be obtained in a range from 1.5 to 12mm (1/16 to 1/2in) thick and with either one or two smooth faces.

CHIPBOARD

This is made from the waste material produced in saw mills, forest thinnings and round wood. It is possible to make it from hardwood, but it is more commonly produced from softwood. It is made so that there are large particles in the core and smaller particles on the face. This is achieved by a screening process or air flotation. The various grades of chipboard stem from varying the particle size, the amount of glue used and the pressure applied during manufacturing.

Single layered chipboard consists of particles that are evenly distributed across the material and are the same size. Coarse surfaces are used for veneering or plastic laminates. *Twice layered chipboard* is produced by partly

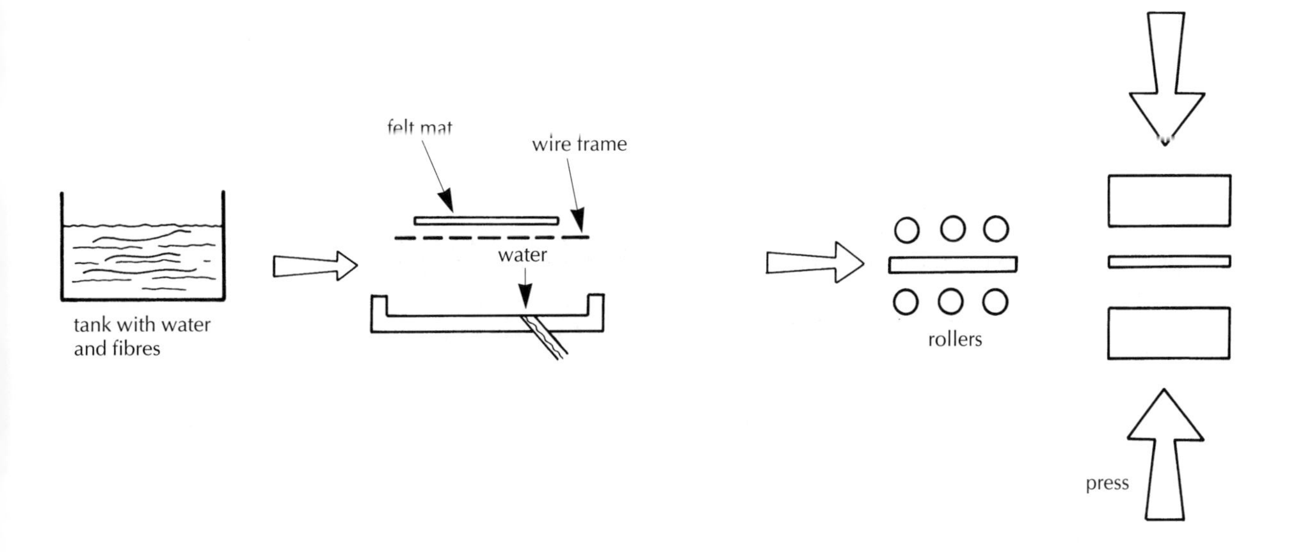

Hardboard manufacture.

sandwiching a coarse particle core. It has a smooth surface and is suitable for most finishes. *Graded density chipboard* is produced with a gradual transition from the coarse to the fine particles. The cheapest of all chipboards is *melamine faced chipboard,* which has a facing of thin melamine foil.

LAMINATED BOARDS

Plywood is made from rotary cut veneers. The layers are called laminates or construction veneers, and their grain direction alternates at right angles between the layers. Types of plywood with different qualities can be produced by varying the amount of glue used in the manufacturing process and using laminates of differing thickness, which leads to differing strengths and appearances. The various types of plywood serve many different purposes – from toys to marine furniture – and the laminates can be cut from a wide selection of timbers.

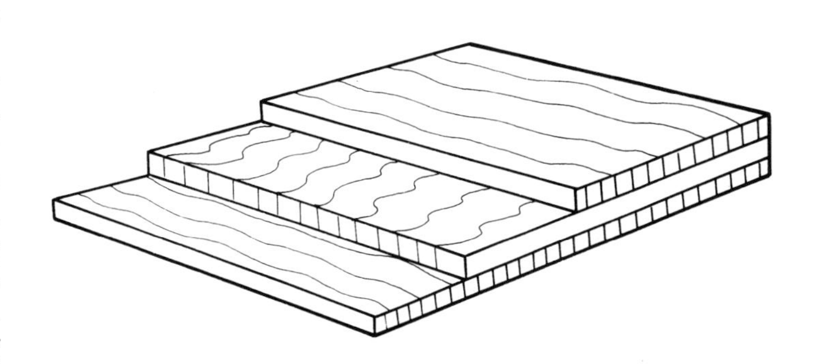

Plywood layers.

Blockboard is a type of plywood where the core is produced from solid timber. The strips that make up this plywood fall between 7 and 30mm (¼ and 1⅛in) in thickness. *Laminboard* is a type of plywood with the core made from strips of wood with a thickness of no more than 7mm (¼in).

Medium-density fibreboard is a wood-based sheeted material consisting of wood fibres bonded together using a synthetic resin

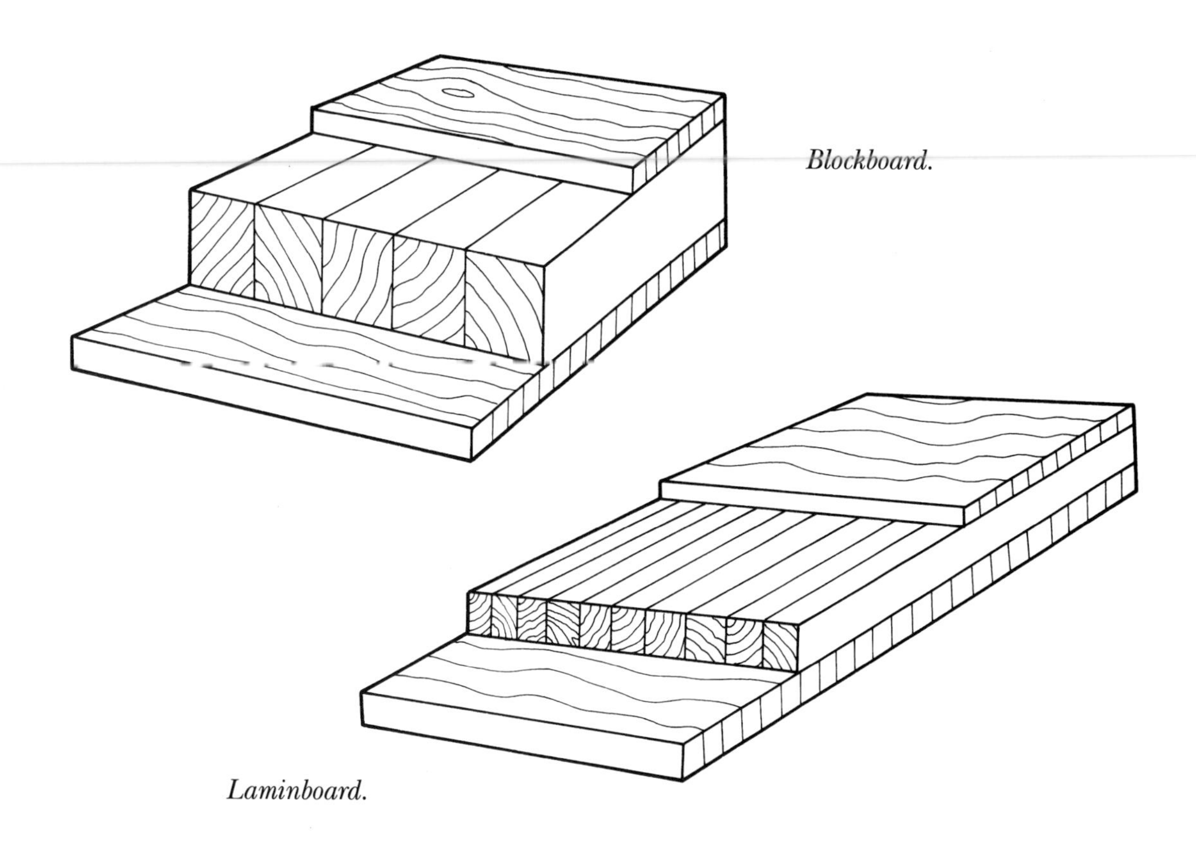

Blockboard.

Laminboard.

adhesive. Its thickness can range from 3.2 to 50mm (⅛ to 2in). It has useful applications in intricate machining operations due to the fact that its fibres are densely packed and uniform. However, it produces a fire dust hazard when machined and it tends to be heavy.

ADHESIVES

Until relatively recent times the woodworker was wholly reliant upon glues made from animal skin and bones. The effects of heat and moisture sometimes led to these glues weakening and the joints failing. The range of synthetic glues currently available overcomes this difficulty to a large extent. Modern day glues fall into three main categories: animal glues, vegetable glues and mineral glues.

ANIMAL GLUES

The most common glue of this type is scotch glue, which can be bought in the form of a hard cake that has to be broken and soaked for approximately eight hours, then heated in a glue kettle. Alternatively, it can be purchased in granules (easier to dissolve, require less soaking time). Never allow the temperature in the glue kettle to exceed 65°C (150°F) and prepare only the amount required for each job as much of the strength is lost if the glue is heated over a long period of time. Avoid the temptation to re-heat the glue. Apply it hot and let it set through the cooling process.

VEGETABLE GLUES

These are glues such as rubber latex, solvent-based rubber contact adhesives, rye flour, soya bean flour and natural resins. These vegetable glues are used only minimally in furniture making, largely because they have poor moisture resistance and so are vulnerable in humid conditions. They do, however, recover much of their strength when they dry out. Their main applications come in the manufacture of softwood panelling and plywood. They are also extensively used in leather work (fixing down leather desk tops, for example) as they prevent the leather from being damaged as it hardens. For plastic laminates it is usual to use a rubber-based impact adhesive.

MINERAL GLUES

Much to the dismay of the traditionalist, many natural glues have been superseded by oil-based synthetic resins. These glues have major advantages over their natural counterparts – they are stronger and more resistant to heat and moisture.

A well-established synthetic glue is *phenolic formaldehyde*. It comes in two forms – hot setting and acid catalyst. The former is a spray dried powder, activated when mixed with water. It produces a hard film that is extremely resistant to the effects of weather and even boiling water. The latter is a high viscosity resin and a liquid acid catalyst and sets by condensation. Be careful as it will remain healthy at a temperature of 30°C (86°F) for only 15 to 20 minutes. This adhesive is heavily resistant to boiling water and the effects of weathering, even at extremes.

Resorcinol formaldehyde comes in a powdered form and can be diluted with either water or alcohol. It will remain usable for about a year if the temperature is maintained at about 20°C (68°F). Store it in a dry place. It sets to a reddish brown colour and is completely weatherproof and water resistant. It is easier to use if bought in liquid form and is a very good general purpose adhesive.

Polyvinyl acetate does not require a hardening agent. It comes in a creamy white water-based emulsion that is ready to use. Its use is limited by its low resistance to moisture and humidity. It is an invisible glue that can only be used for dry interior applications. A solvent-based polyvinyl acetate is on the market and has better resistance to moisture.

Urea formaldehyde can be obtained as a two-part liquid resin or as a combination resin and hardener in a powdered form that is activated by the addition of water. In the first form the hardener is applied to one part of the joint and the resin to the other and it sets when the whole joint is clamped.

Epoxy resin glues come in solid and liquid forms and are useful for work on a variety of materials, including metal, wood and glass. In liquid form they are not especially long lasting. It is possible to purchase synthetic resins that set only when the resin comes into firm contact with the hardening agent. These are useful in some circumstances as one surface can be smeared with the hardener and one with the resin and the joint will be glued as the two come into contact. These are known as separate application glues, while those that are already mixed in the correct proportions are known as mixed application glues. Synthetic resins can be purchased with mixed resin or as hardener in a powdered form that is activated by water.

PREPARING THE GLUE

Most glues must be heated before they are ready to use. One method is to use an electric glue kettle, the other is to use a glue pot.

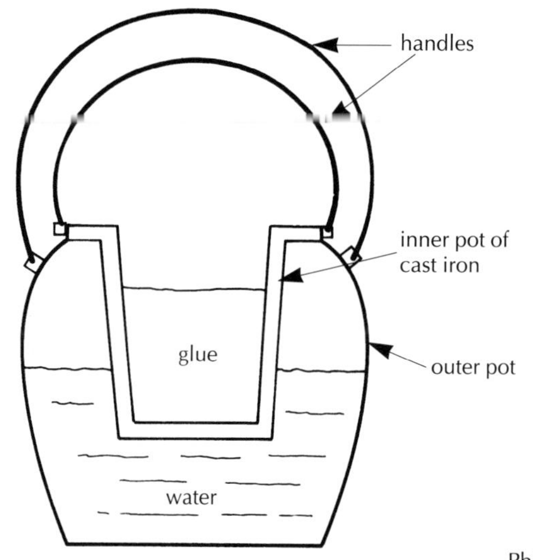

Glue kettle.

A glue pot is usually made from cast iron. Make sure that the inner pot is free from rust as the oxidation may discolour the outer glue pot, which is also made from cast iron. It contains water, which stops the glue from over-heating. To prepare the glue, place pearl glue in the inner pot up to about half full, pour in the water to roughly 10mm (⅜in) above the level of the pearl and let it soak through. It should then turn to an applicable jelly.

WOODSCREWS

Woodscrews have two main uses – for joining wood to wood and for attaching fittings to wood (such as hinges, locks and handles). They have three main types of head. *Countersunk* heads have a flat head sunk flush with the timber and either slotted or with a cross head. *Round head* screws are normally used to fix a flat sheet to wood and are again either slot- or cross headed. *Raised screws* are part countersunk and part rounded and have a domed top or slot head.

In conventional wood screws about 60 per cent of

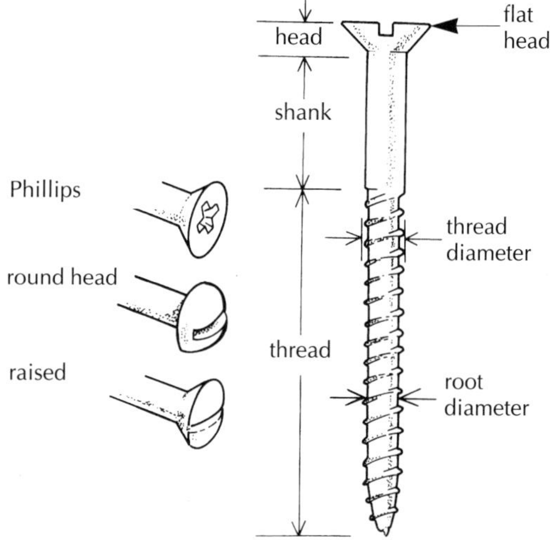

Dimensions and types of screw.

Length measurements for flat, raised and round head screws.

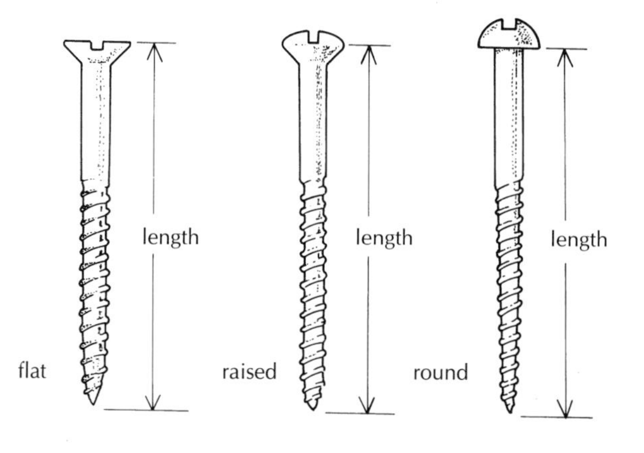

TIP

It is a good idea to ease the passage of a screw by counter-boring a hole into the timber. With standard screws this will require two drills, one the size of the shank and one the core size of the thread. With machine screws just one hole is required, for the core size of the thread. A dab of wax on the screw also helps.

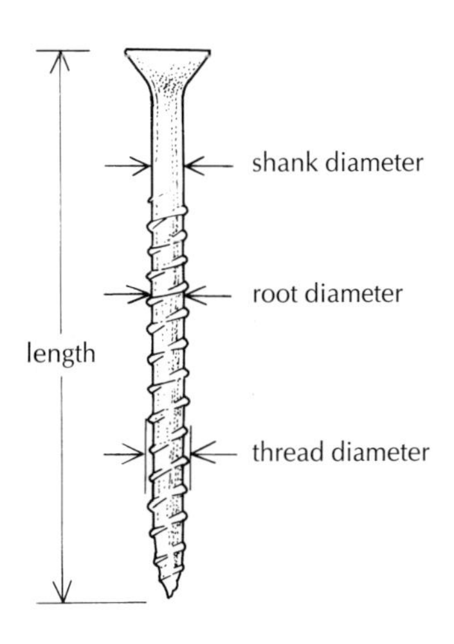

Flat head screw

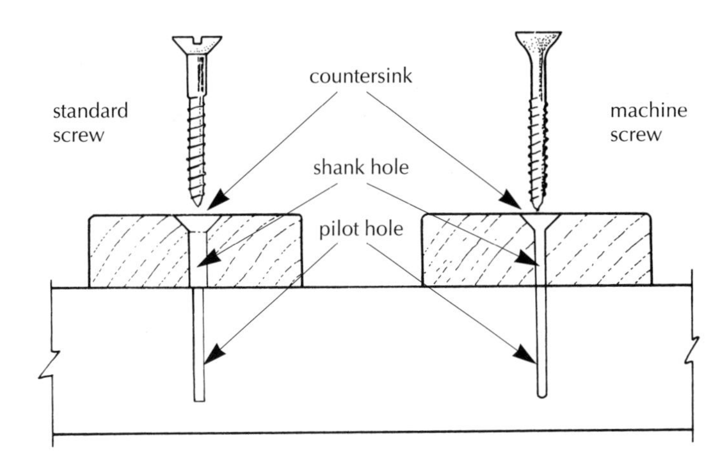

Countersinking screws.

the length has a thread. The shank acts as a dowel.

A new type of screw is the *twin fast.* This has a twin thread instead of the conventional single thread, and is very good for fixing synthetic boards such as chipboard. More of its length has a thread and the shank is smaller in diameter than on a conventional screw, so there is less risk of splitting the timber.

CHAPTER **four**

Planing and Sawing

Of all the techniques and skills that have to be mastered during a working lifetime, perhaps the most fundamental techniques are planing and sawing. There will be few jobs for which these processes are not the natural starting point – in particular, correct plane operation is absolutely vital in many areas of timber work.

PLANING

The development of a good planing technique is crucial to a cabinet maker: it involves appreciation of grip, stance, and arm and shoulder movement. Adapting to the range of planes used over the years should merely require slight modifications to the basic technique.

The most fundamental consideration is that the cutting edge remains straight and level in nearly all circumstances. In this chapter there is an outline description of the basic requirements of a good planing technique, though allowance should be made for the idiosyncrasies of the individual. It is important to remember that the plane must be seen as the natural extension of its user and this will not be achieved if the working position seems artificial and uncomfortable.

HOW TO USE THE BENCH PLANE

Under most circumstances work should normally be planed on the workbench using a bench stop or dog, and only occasionally when secured in the vice. There are times when it may be necessary to make use of the vice to hold the workpiece in place, but these occasions should be kept to a minimum. There are obvious

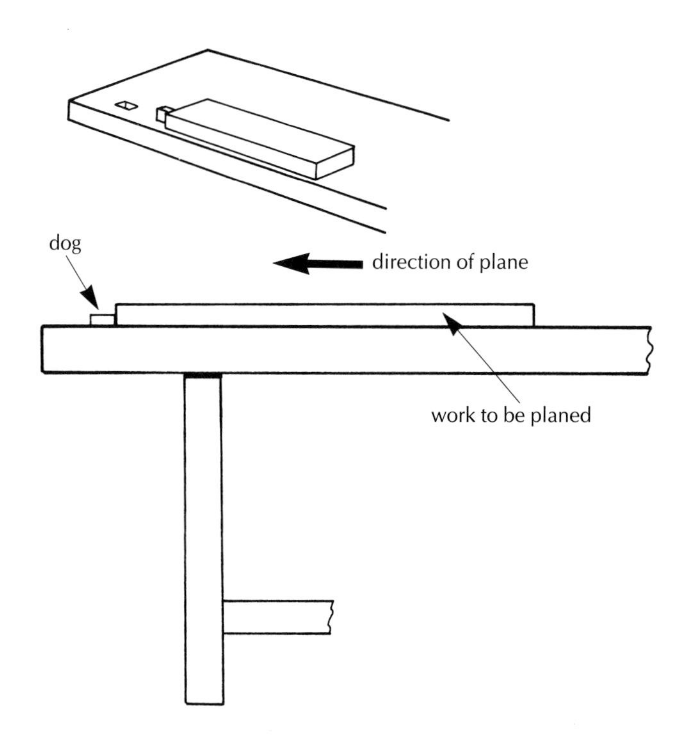

Using the bench plane.

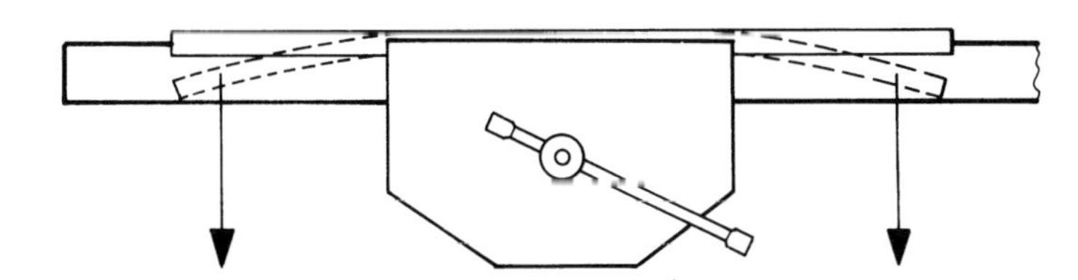

Problems when using the vice.

and valid reasons why the vice is seldom used: there is the real danger of slippage and the primary concern that thin sectioned timber may deflect, making it impossible to achieve a flat and true face.

STANCE

The stance adopted when planing is important, yet it is often neglected by the novice. A correct and comfortable stance will allow the individual to create a natural and easy flow of pressure from the upper body smoothly into the action of the plane. If done correctly it will prevent the jerky and erratic movements that can scar timber unnecessarily.

In general, stand so that the movement of the arm and shoulder is free but controlled. Remain firm in order that the plane can glide across the workpiece. To achieve this position, attention needs to be paid to the position of the feet. The left foot should be directly in line with the workbench, while the right foot is placed at right angles to it. With practice this will become second nature.

How to hold the bench plane – note the directional finger.

HOLDING THE PLANE

It is important that the plane should be held firmly and correctly and become a comfortable extension to the adopted stance. It is at this point that the pressure emanating from the hands, arms and shoulders is applied, to create the smooth cutting action that is required. The left hand should be holding the knob and the right hand the handle, with the index finger pointing towards the knob.

To plane the surface flat, particularly on rough or preliminary work, use the jack plane. For planing long timber the try plane can be used to achieve a true levelling, then a finely set smoothing plane for

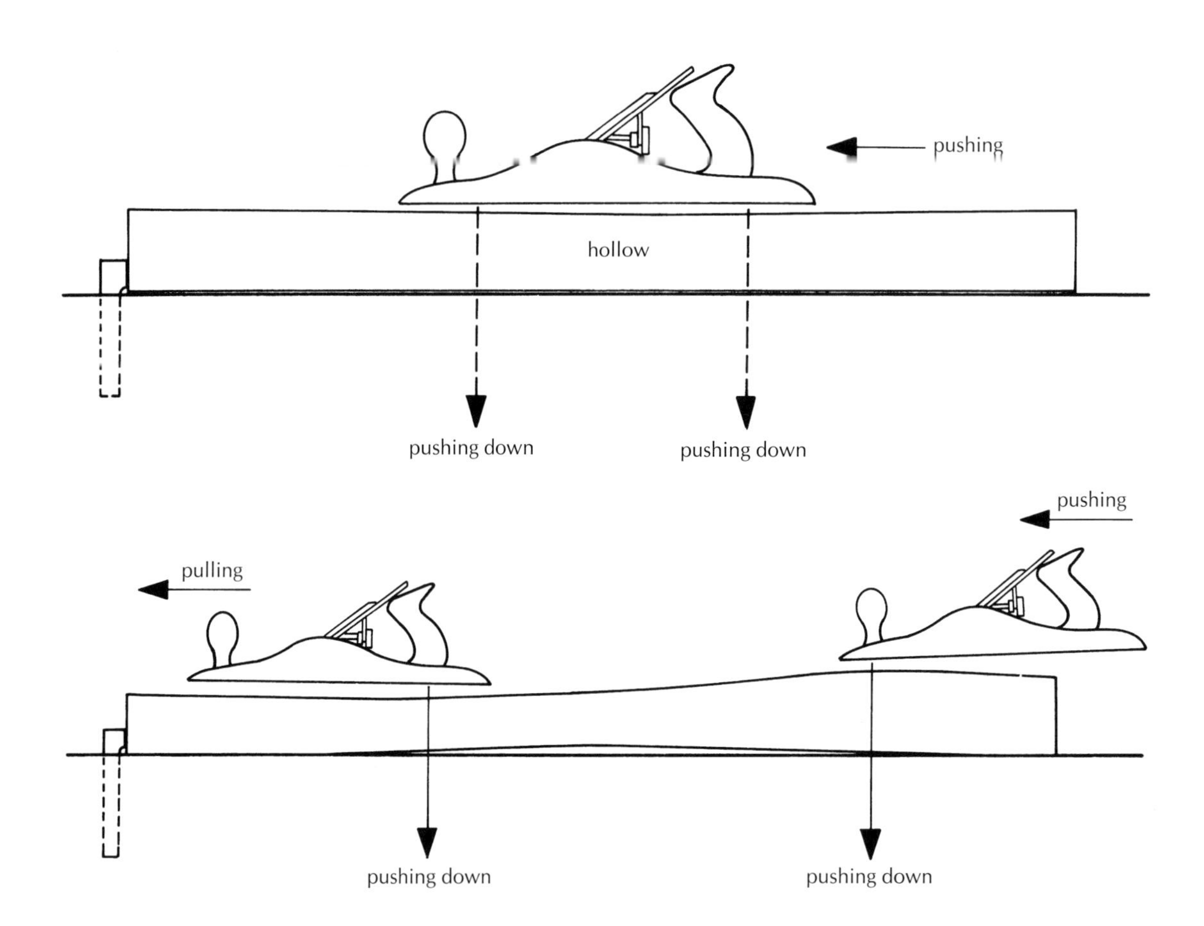

Planing irregular wood.

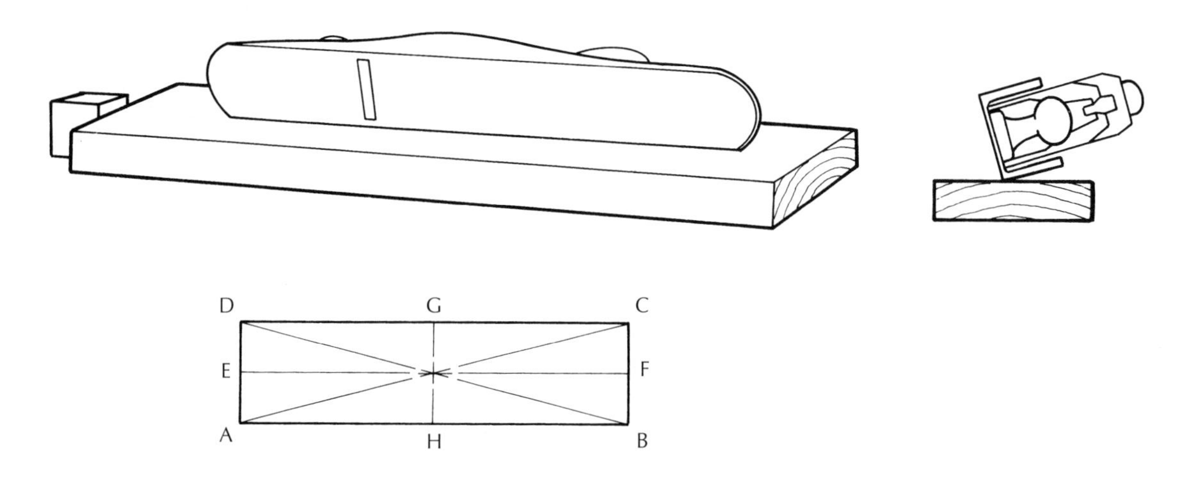

Check for flatness along lines (E–F, D–B, A–C, G–H).

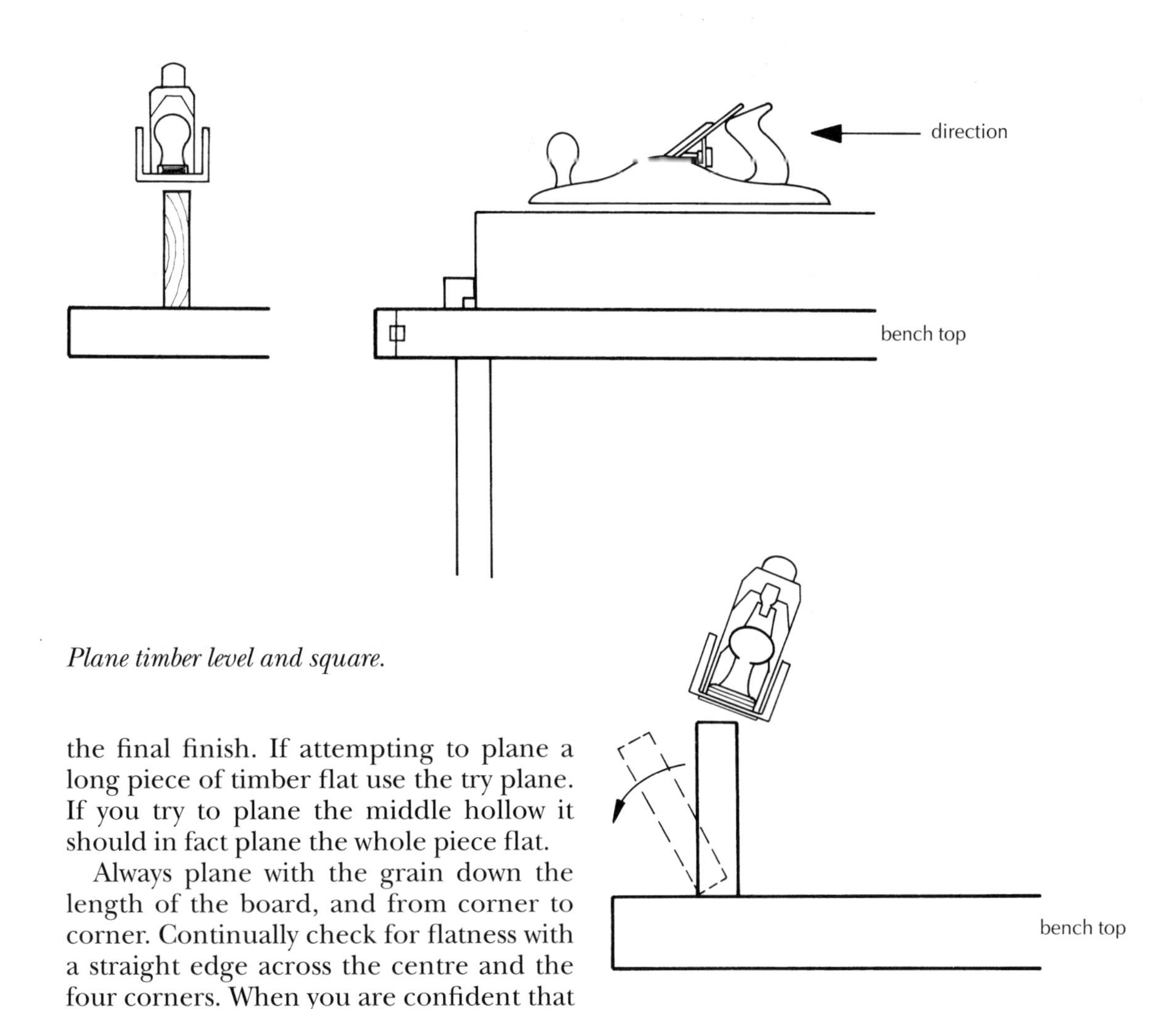

Plane timber level and square.

Timber falls if not square.

the final finish. If attempting to plane a long piece of timber flat use the try plane. If you try to plane the middle hollow it should in fact plane the whole piece flat.

Always plane with the grain down the length of the board, and from corner to corner. Continually check for flatness with a straight edge across the centre and the four corners. When you are confident that the surface is flat, mark with a pencil on the face side.

EDGING

Once the surface is planed true the next step is to plane the leading edge true and straight across its length and at right angles to the face. It is a difficult operation that needs great care for it is much harder to plane the edge precisely true to the face than it is to simply plane a straight edge. Complications can arise if the stance is incorrect or the planing action is not smooth or the workbench itself is not perfectly true and level.

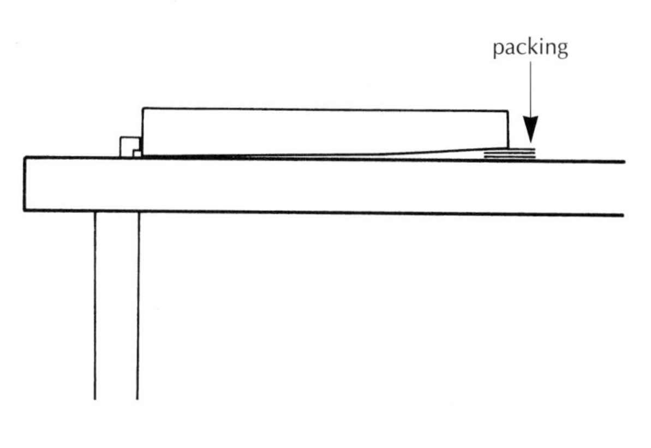

Pack timber level.

TIP

For planing the end grain choose a metal plane. The fibres at the end may break out and splinter as the plane passes over the edge, but there are a number of ways of preventing this from happening:

- Plane the timber from both ends to the centre.

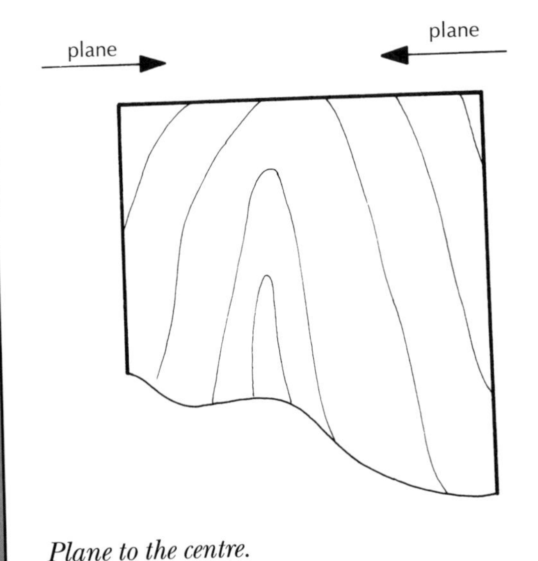

Plane to the centre.

plane

waste

Leave some waste.

- Leave some waste on the width of the board and cut a nullifying chamber on the waste.
- Clamp a chambered waste block onto the end of the board.
- Make use of a shooting board. This is particularly useful if the edges of the timber are too thin to accept the sole of the plane.

If the timber falls over it is usually because the plane is not being held perfectly level. The timber will also fall if it is not being planed square. If the timber is not flat, pack it level with shavings, paper or card to stop the timber rocking on the workbench. Whatever the difficulties associated with accurate and precise edging, it is a fundamental technique that has to be competently mastered.

SAWING

There are only two basic types of hand saw: the cross-cut saw, which is used for cutting across the grain, and the rip saw, which is used for cutting with the grain. It is sometimes felt that sawing is such a common, everyday activity that it is somehow easier than other seemingly more complex cabinetry activities. However, this is

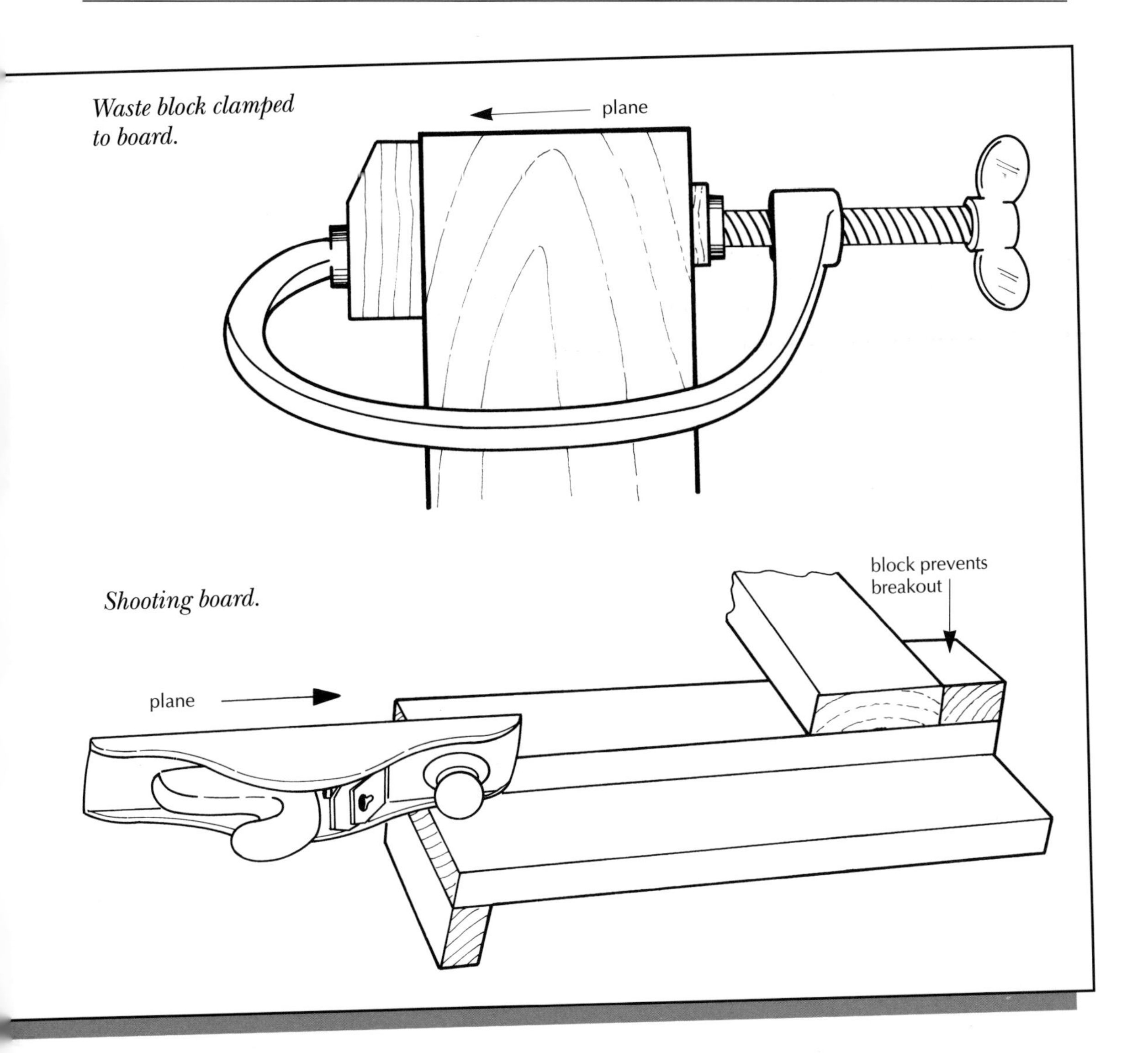

Waste block clamped to board.

Shooting board.

not the case. Great care and attention is needed when sawing. Careless and avoidable mistakes can be costly in time and money, so it is important to consider the fundamental principles.

USING THE CROSS-CUT SAW

Starting the cut correctly is absolutely critical. Place the thumb of the non-sawing hand along the mark and rest the blade against the thumb. Never start with a down stroke as this cannot only be dangerous for the thumb and fingers, but can also cause the wood to be damaged if the saw jumps – a couple of short, controlled up strokes should suffice. The most important thing to bear in mind at this point is the grip on the handle of the saw. The index finger of the right hand should be placed alongside the handle and pointed in the direction and at the required angle of the cut. This

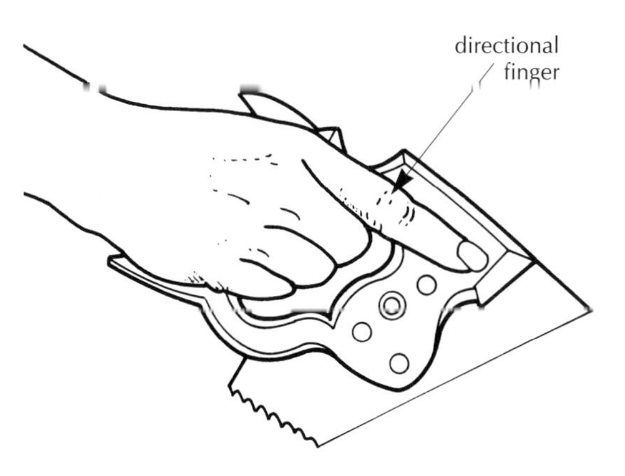

Gripping the saw.

to the stance have to be made to meet individual circumstances. If cutting the end of a board that is resting on stools, support the waste end with the left hand and hold it in a position that will allow it to assist in preventing the saw from binding. If a long piece of wood requires cutting in the middle it is often better to have a colleague hold the timber at one side, always paying attention to the angle of the timber so that the saw does not become trapped.

USING THE RIP SAW

The rip saw is gradually being replaced in most workshops by powered saws. But there will remain occasions when the hand saw is necessary and it will remain in common

will allow the blade to be guided accurately and correctly in the right direction and acts rather like the sight on the barrel of a gun.

The sawing motion should be as natural as possible and should engage the arms, legs, shoulders and hands in unison. Let the weight and the natural movements of the saw lead it into the wood. Try not to force the saw through the timber with too much pressure as this will make it difficult to control. But try not to use short, abrupt cutting strokes, which will produce similar results. Instead draw the saw slowly backwards and then press it slowly and gently forwards in long flowing strokes.

The stance taken up by the cabinet maker will be determined by the length of wood to be cut and modifications

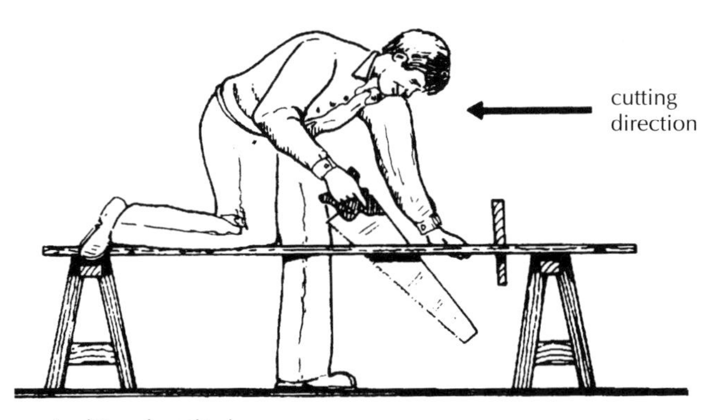

Sawing methods.

usage for some time to come. It is therefore useful to know how best it can be applied.

The grip and cutting action are essentially the same as for the cross-cut saw, but there are some adaptations that can be useful. In the traditional way the timber is placed across two trestles with the right knee pinning it firmly down and the left leg and arm providing the support necessary to allow for the cutting action of the right arm. The problem with this method is physical rather than technical – the stance can be uncomfortable and can lead to fatigue if it is maintained for any length of time. An alternative method is called the over-arm method and has a number of advantages. It is less tiring because the cabinet maker sits astride the timber, using body weight to provide support and stability for the wood. Both arms are used simultaneously, with the left hand on the front of the handle and the right hand behind. The workload is more evenly distributed and this results in a reduction of fatigue levels. It is also easier to see what is happening to the angle of the blade and to assess the accuracy of the cut.

To achieve the best position, put the timber across two trestles placed somewhat closer together than for the conventional method. Sit slightly to the front of the rear trestle and saw along the timber. Resist the temptation to follow the cut; pull the timber backwards under you to continue the sawing action. When this method is employed the saw is used at a slight angle rather than at 45° as in the conventional method.

TENON SAW

The tenon saw, as the name suggests, is largely used for cutting tenons and consequently is most commonly employed in conjunction with a bench hook to hold the timber in place. When undertaking this process hook the bench hook against the front of the workbench and hold the timber against the top block of the hook with the left hand. Failure to hold the timber securely will lead to errors and a subsequent loss of accuracy, time and materials. Start the cut with the same up stroke methods described for the cross-cut saw and commence the sawing action with the blade at about 30°. As the process continues the blade is gradually inclined until it lies horizontally.

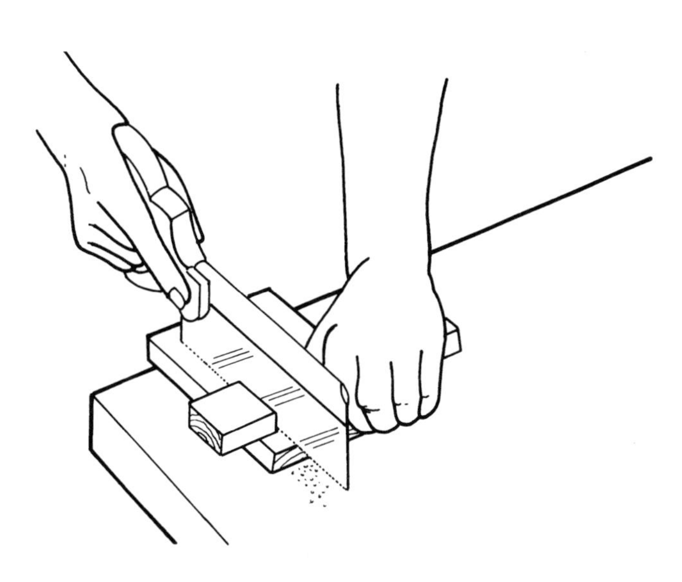

Using the tenon saw and bench hook.

CHAPTER 

Edge Joints

Edge joints, otherwise known as butt joints, are designed to joint narrow widths such as facings. There are a number of named joints that fall into this category, with the nature of the piece being worked on and the demands placed upon it determining which type of joint is constructed. Many of the most useful joints are plain and simple glued joints, but others need additional refinement to meet specific requirements, such as providing for extra support.

TIP

The greater the area of timber that is available to house the glue, the stronger the joint will generally be, depending on the type and strength of the glue applied. It is possible to reinforce joints that need to support comparatively heavy weights by using screws, tongues or dowels.

To ensure adequate bonding the edges that are to be joined must be planed straight and square using a try plane, then glued together. This is not as easy as it sounds. Be careful not to underestimate the difficulties that can be encountered in this process by imagining that it is simply a case of 'sticking' together two pieces of wood. Glued jointing, particularly on long pieces of timber, requires a high degree of skill and patience. If the edges are not planed precisely to fit snugly together or they are not perfectly straight, the result can be a cupped board or a weak joint.

Two basic methods can be employed to produce a glued joint. In *rubbed joints* two straight joints are glued and immediately rubbed together, the natural ingestion of the glue providing the tight jointing. *Cramped joints* (or clamped joints) are planed slightly hollow, glued with cold setting resin and firmly joined with a cramp or clamp.

GLUED EDGE JOINTS

Glued butt joints are the simplest form of glued joint and are normally used when two or more boards are to be joined together.

Rubbed joint.

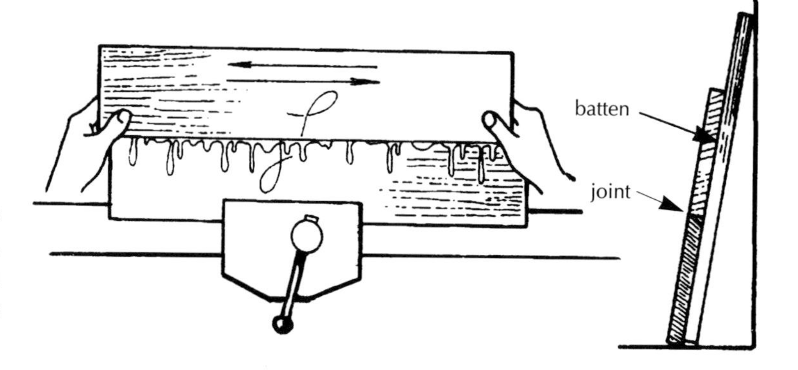

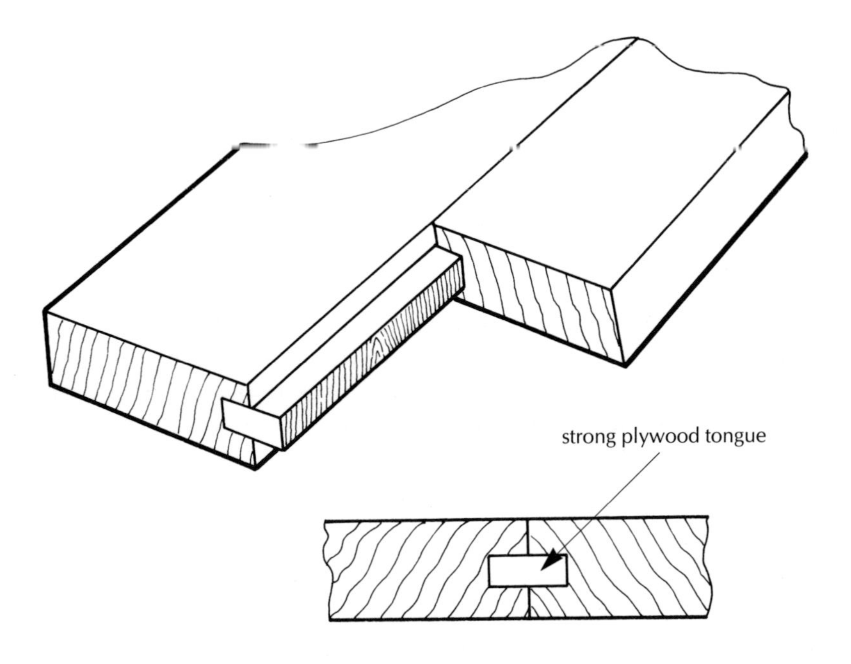

Loose tongued joint.

LOOSE TONGUED JOINTS

A tongued joint is generally to be preferred to a plain glued butt joint because it gives added strength, especially in such items as unsupported table tops. The main reason for this added strength is the greater gluing area that this joint provides.

Using a plough plane router, a groove of uniform depth and width is cut into the board to be jointed – which has had its edge planed straight and square. The tongues were traditionally made from a thin piece of cross-grain timber planed up parallel and square. Nowadays, it is common practice to make the tongues of good quality birch, or ideally plywood of a suitable thickness. To ensure a good joint the tongue should slip closely into the groove, but should not be too tight. The tongue should be narrower than the combined width of the grooves.

Loose tongued joints are most effective but require a degree of skill and practice, and they can be somewhat cumbersome. The consequences of small errors with tongues cut too long, too short or too thick can be costly in wasted time on repairs and discarded material. In recent years, joints that are a less intricate variation on this theme have become more popular, and are sometimes considered more cost effective as they are quick and relatively simple to construct.

The *biscuit joint* is cut using a purpose-made machine with a small circular saw blade. It resembles a biscuit in shape, hence its name. The *'F' joint* is cut using a spindle moulder or router with the appropriate cutter or block. For a *tongue-and-groove joint* the groove is cut in the edge of one board, as with the loose tongue. The tongue is formed on the edge of the other board with a plane or router, or a spindle moulder with a tonguing cutter. If three or more boards are to be jointed, the intermediate boards will have one edge grooved and the opposite edge tongued.

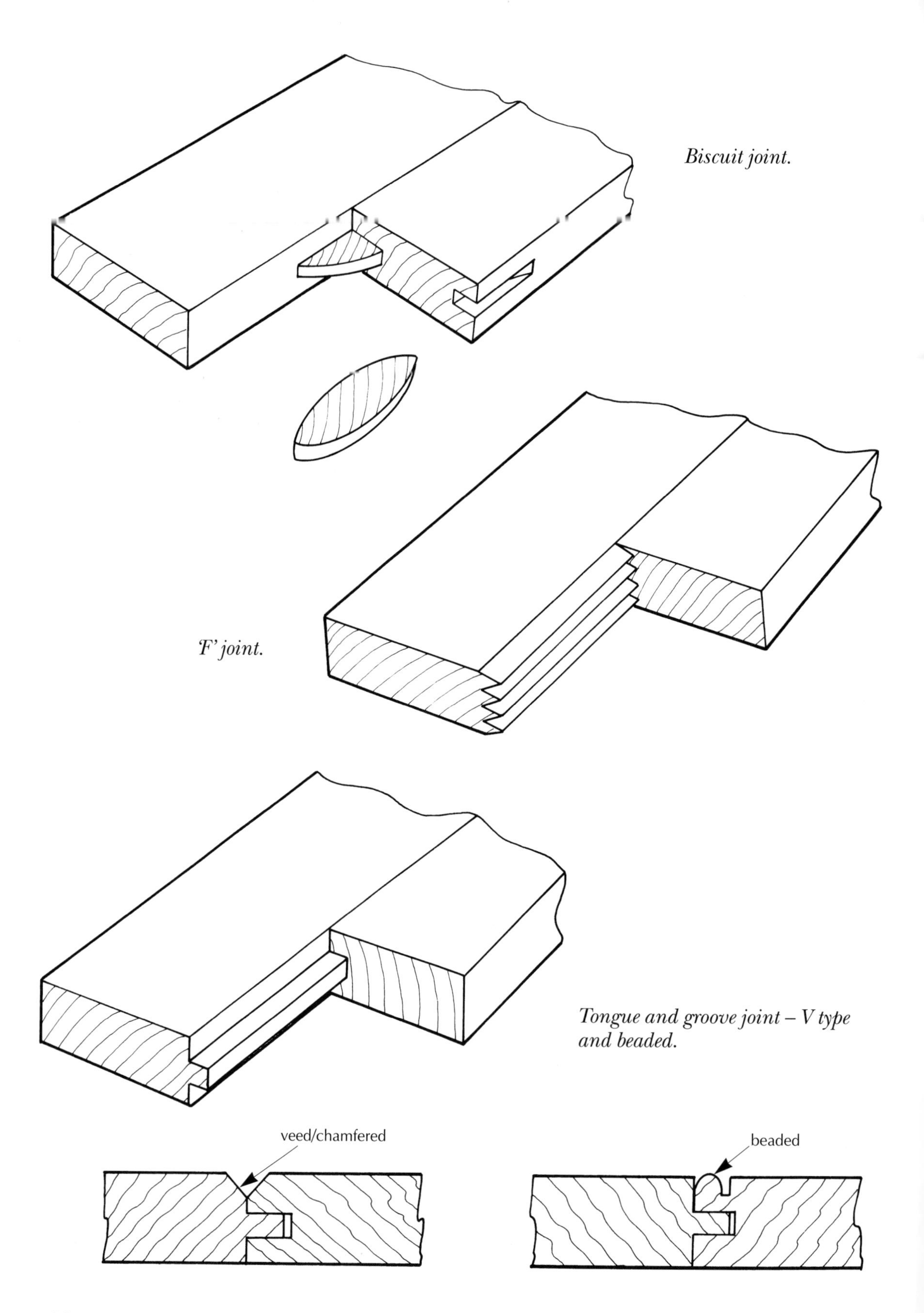

Biscuit joint.

'F' joint.

Tongue and groove joint – V type and beaded.

When the screw is turned 45° the countersink of the bevel pulls together the two compatible edges. The edges can then be glued and clamped.

STRENGTHENING JOINTED BOARDS

Jointed boards are used in cabinet making and you may come across them when restoring the doors of some older types of furniture. Batten joints are jointed boards strengthened with the aid of battens screwed onto the back. To allow for the movement of the doors the screws are slotted. Dovetail battens can be made either parallel or wedged and have the advantage of not requiring glue or screws to hold them in place. Buttoned battens need the use of a solid screw.

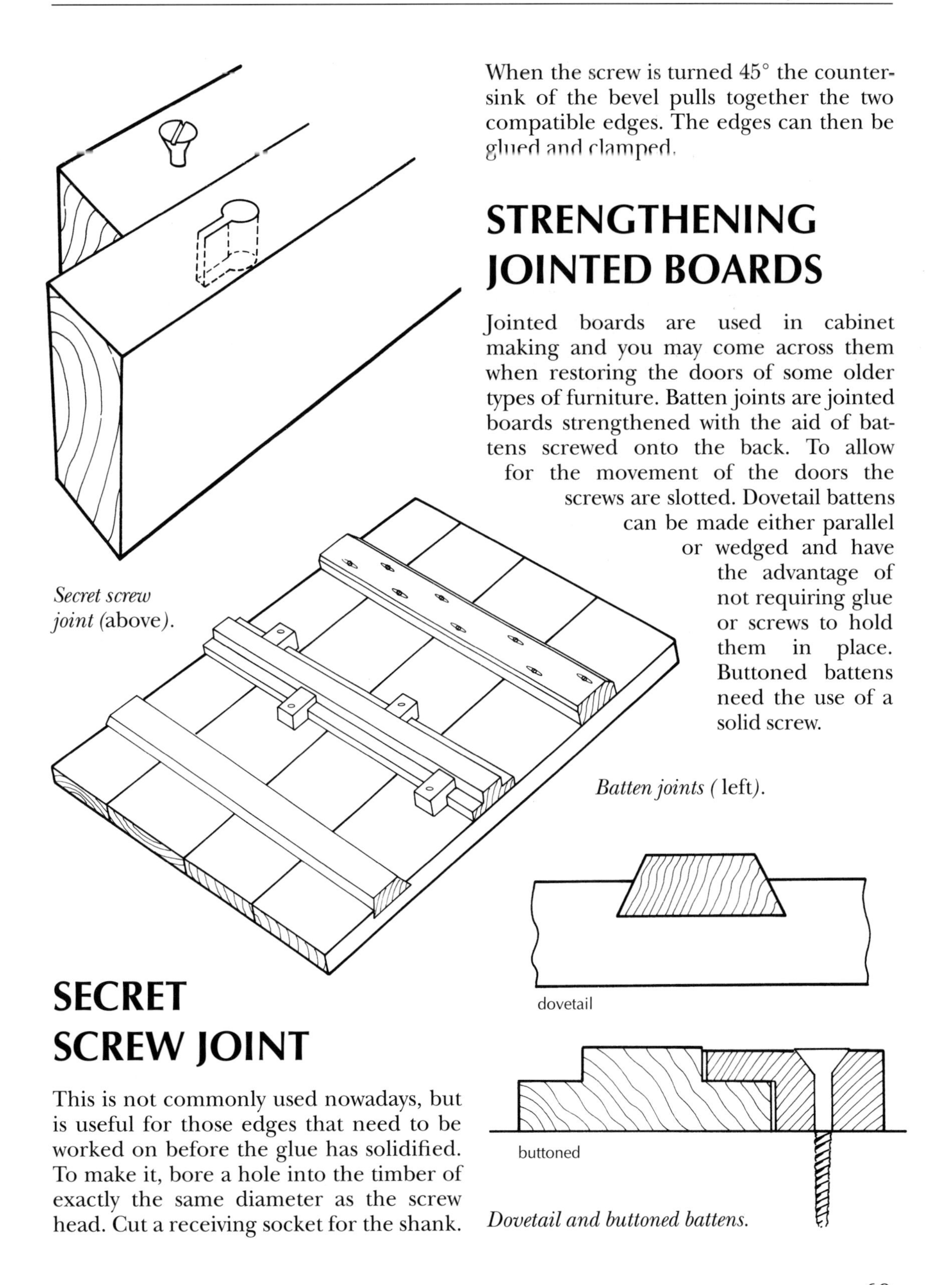

Secret screw joint (above).

Batten joints (left).

Dovetail and buttoned battens.

SECRET SCREW JOINT

This is not commonly used nowadays, but is useful for those edges that need to be worked on before the glue has solidified. To make it, bore a hole into the timber of exactly the same diameter as the screw head. Cut a receiving socket for the shank.

CHAPTER

Mortise and Tenon Joints

The mortise and tenon joint is one of the most universal joints to be found in woodworking. It is popular as it is relatively straightforward to construct, and can be hidden from view or exposed to add a decorative feature. It is flexible and can be adapted and modified to meet a huge number of specialized requirements, while maintaining its simplicity. Much of its popularity stems from its most important quality – its inherent strength. Its sturdiness is largely derived from the relatively large glued area. The mortise is the opening of the joint and the tenon is the tongue that slips smoothly into it like a key into a lock.

Although there are a large variety of forms to choose from, all mortise and tenon joints are constructed along fairly general principles. It would not be possible to give a detailed account of how to make all of them, so an overall description illustrating the production of a haunched mortise and tenon joint will be given by way of example, with a brief commentary on the more useful alternatives.

MORTISE CONSTRUCTION

Carefully set up the mortise gauge by adjusting the spurs to match the width of the chisel. Using a try square mark out the length of the mortise. It is always advisable to leave at least 12mm (½in) spare at the waste end to allow for mishaps and splits. Set a mortise gauge to the correct dimensions and mark the mortise line on the wood. Great care should be taken over this procedure to make sure that the marking is precise.

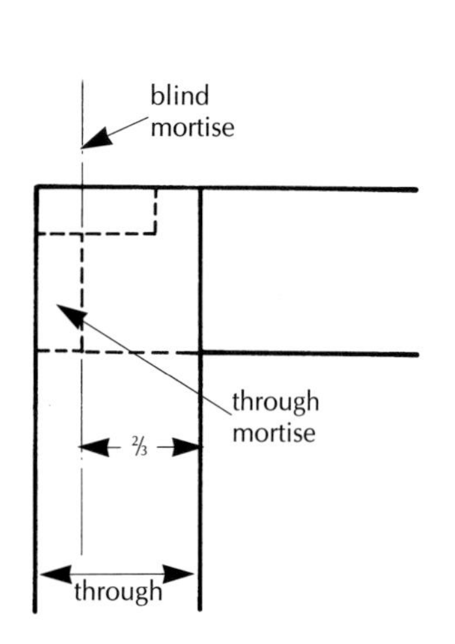

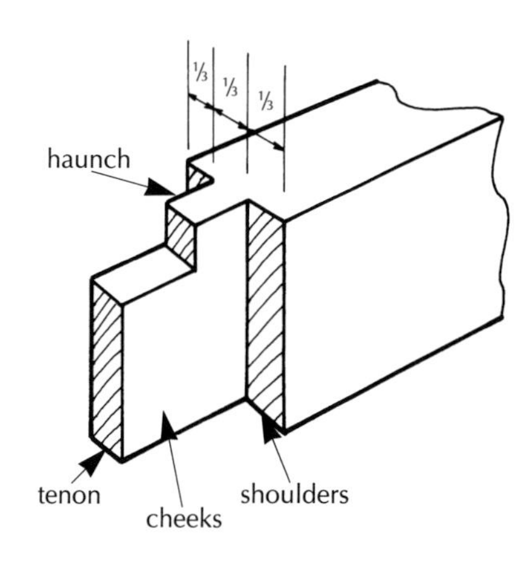

Parts of the mortise and tenon joint.

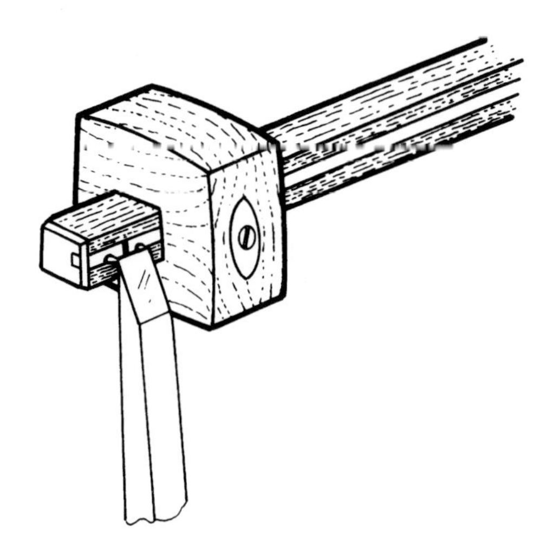

Mortise gauge set to chisel width.

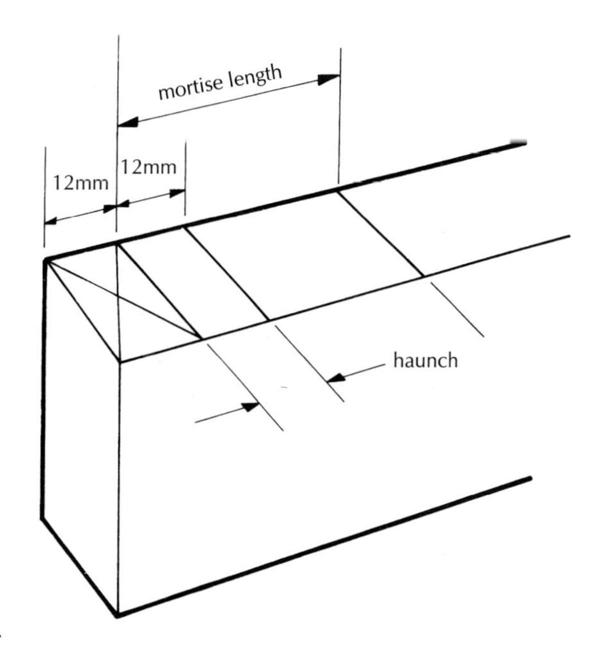

Marking out the mortise.

TIPS

When working on any type of mortise and tenon joint remember that the thickness of the tenon is always equal to one third of the thickness of the material it is cut from. For example, for 25mm (1in) wood the rail would be 8.3mm. Select the nearest chisel to that width using this rough guide:

Rail thickness	*Chisel size*
18mm (¾in)	6mm (¼in)
22mm (⅞in)	8mm (5⁄16in)
25mm (1in)	8mm (5⁄16in)
28mm (1⅛in)	9mm (⅜in)

Always cut the mortise first. The tenon can then be cut to correspond to the opening exhibited by the mortise.

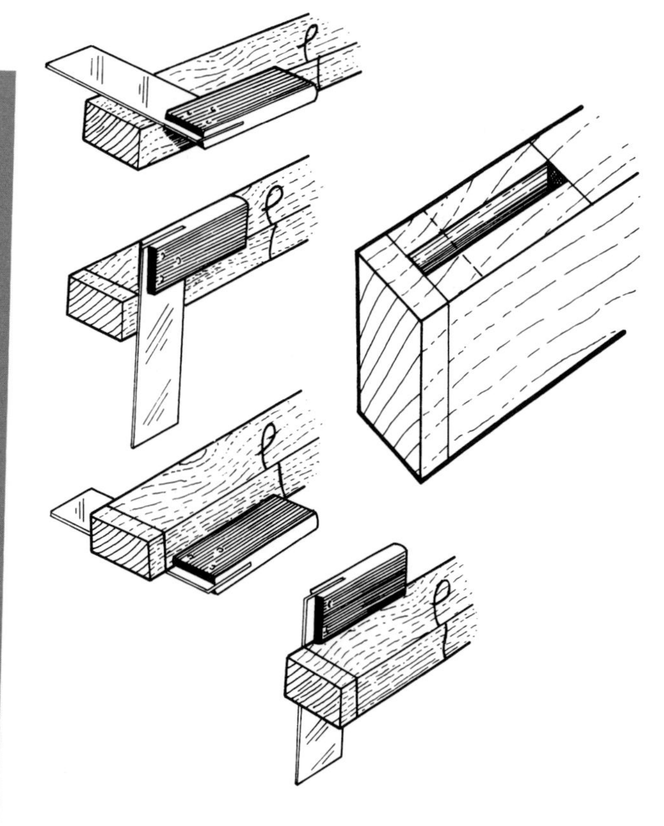

Score a line around using the try square.

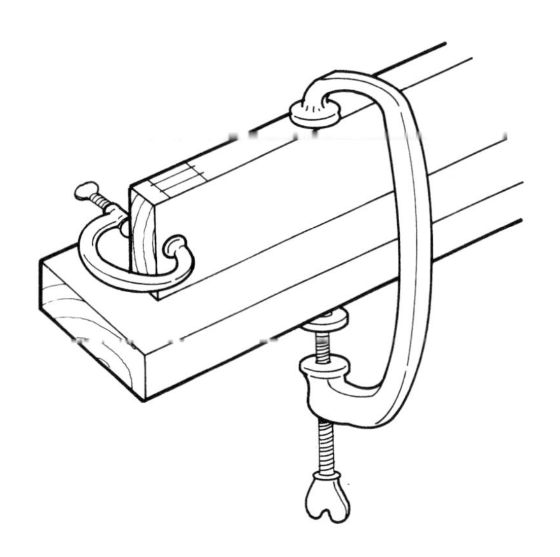

Cramp to the workbench.

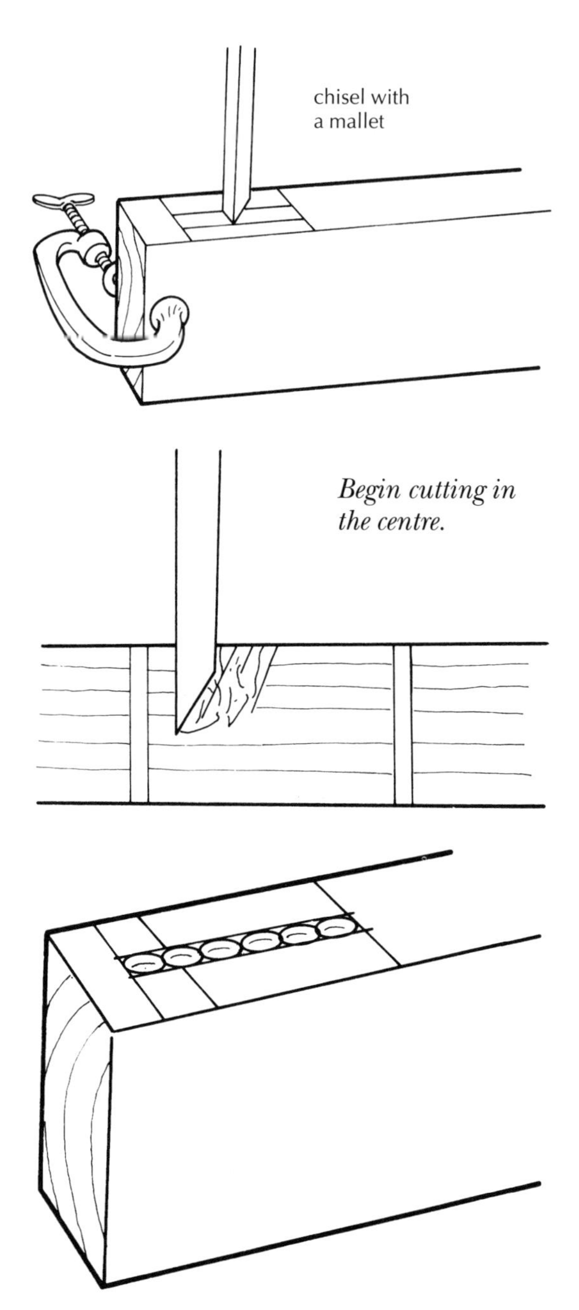

Begin cutting in the centre.

Holes drilled to remove waste.

When you have done this, cramp the work tightly and securely to the workbench. Leave plenty of room to allow yourself to adopt a comfortable working position. Then, starting from the centre and working outwards use the correct mortise chisel to begin cutting into the wood. It may be necessary to use a mallet to provide the necessary power. Always use a wooden mallet – never a hammer – and when striking the head of the chisel use short, controlled hits as overly aggressive strikes may dislodge the timber from the cramps or cause it to split. As you progress studiously, lever out the waste so that you remain well sighted. Continue with the vertical chiselling, repeating this operation at regular intervals. (The waste can be removed by drilling a series of holes to the necessary depth without scarring the sides of the mortise and then chiselling out.) Finally, clean out the sides of the mortise with a sharp chisel.

The biggest problem when cutting out the mortise on a stubbed joint is to determine exactly how deep it should be cut. Insufficient depth will weaken the joint, while a mortise cut too deep runs the risk of splintering the opposite edge. As a general rule it should be cut to about two thirds of the width of the timber into

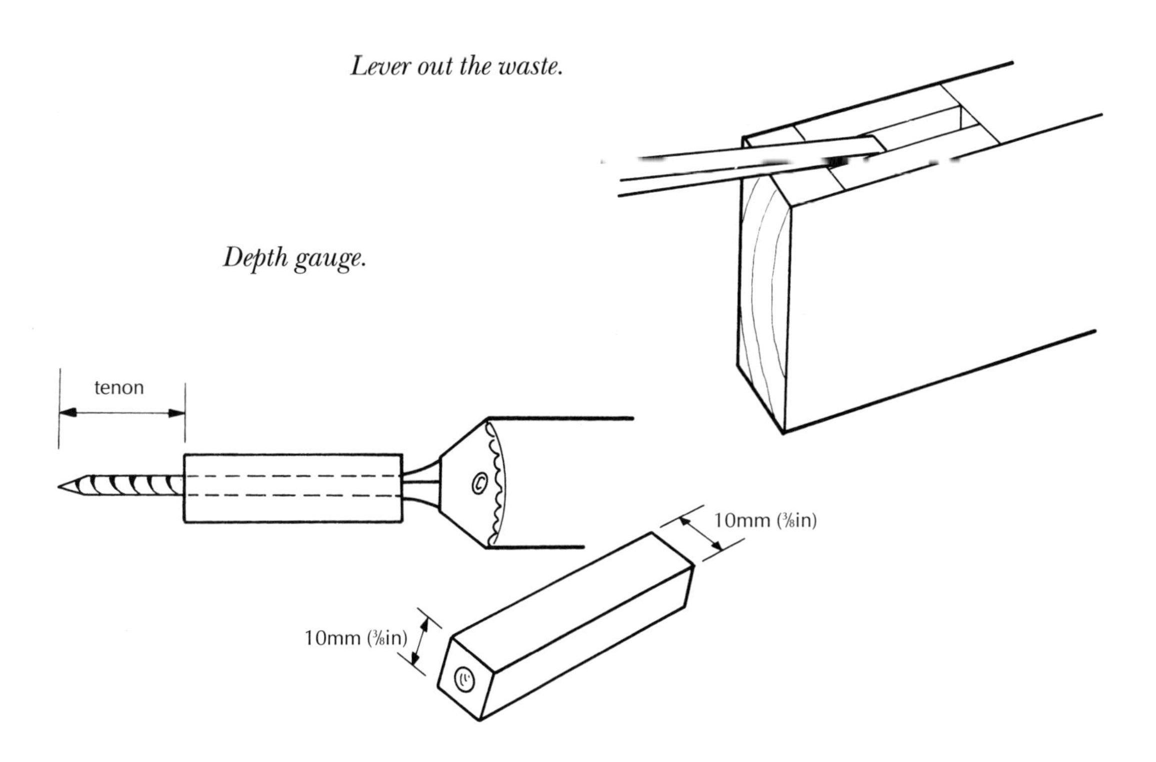

Lever out the waste.

Depth gauge.

which it is cut. Do not go beyond this point as the result could well be splintering. A useful device to employ when measuring the depth of the cut is a basic depth gauge set to the depth required in the mortise. Periodically place it into the cut to gauge the depth.

TENON CONSTRUCTION

Once the mortise has been completed it is now time to construct the tenon. First measure the depth of the mortise and translate this figure to the timber from which the tenon is to be cut. Square the shoulders and then mark out the necessary mortise gauge lines. Three sets of parallel lines will be needed (across the face and on the two edges). Securely place the timber into the workbench vice at around a 40° angle and cut out along the lines with a tenon saw (*see* overleaf). Do the same for the other edge and then place the timber into the vice horizontally and cut down the required depth. Use a bench hook when cutting the shoulders.

If it is to be a haunched joint, mark out the haunches and saw off the waste timber. (The haunches add strength and versatility and can often be seen in the construction of cabinet doors as they can accommodate their swinging motion without any loss of fortitude.) Test the joint by placing the newly cut tenon into the mortise. There should be a very small space between the tip of the tenon and the base of the mortise to allow for excess glue when the piece is finally jointed. If necessary, adjustments can be made using a shoulder plane.

All grooved frameworks require haunches. If there are no grooves it is possible to fit a sloping haunch.

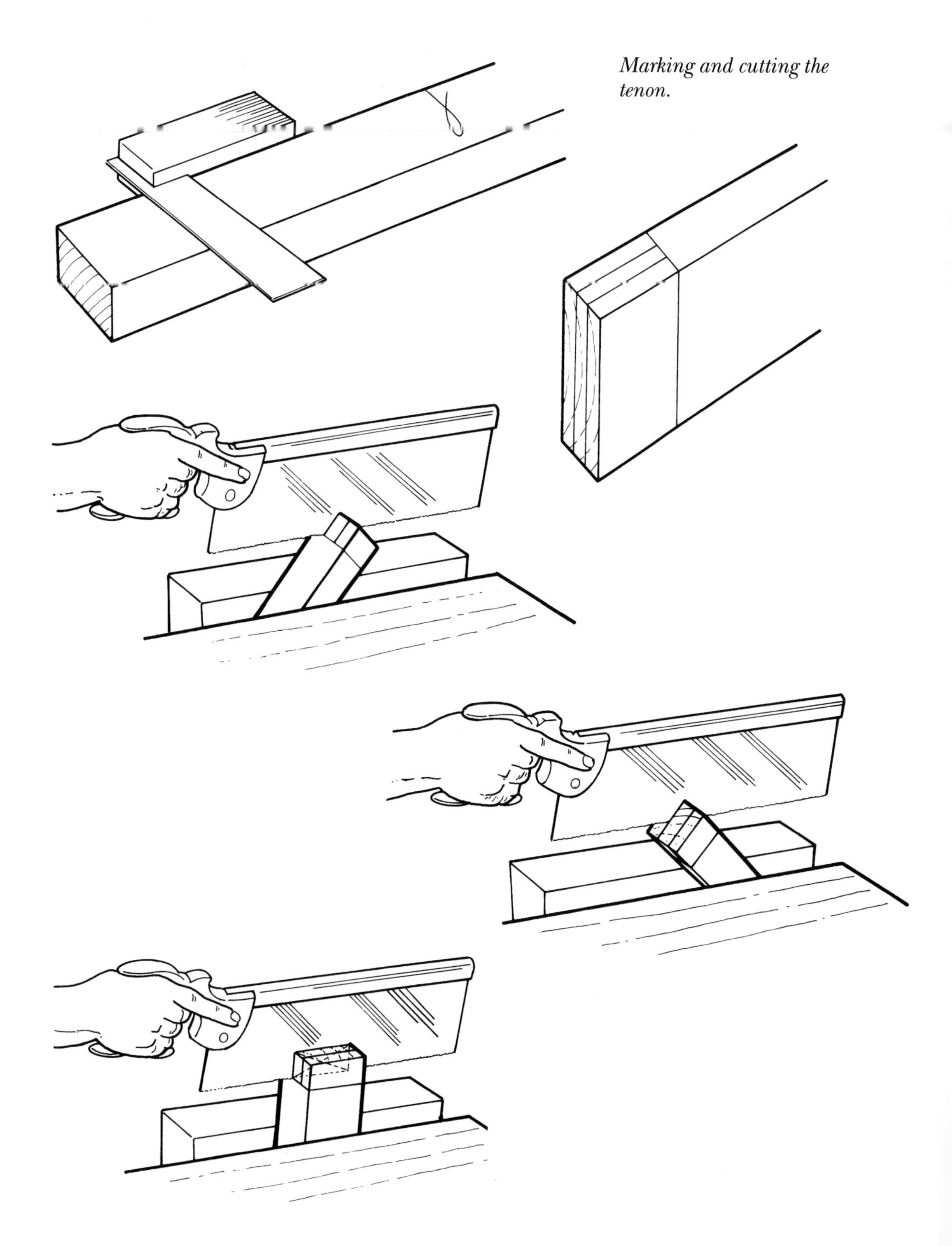

Marking and cutting the tenon.

Cutting a haunch.

Marking a square haunch.

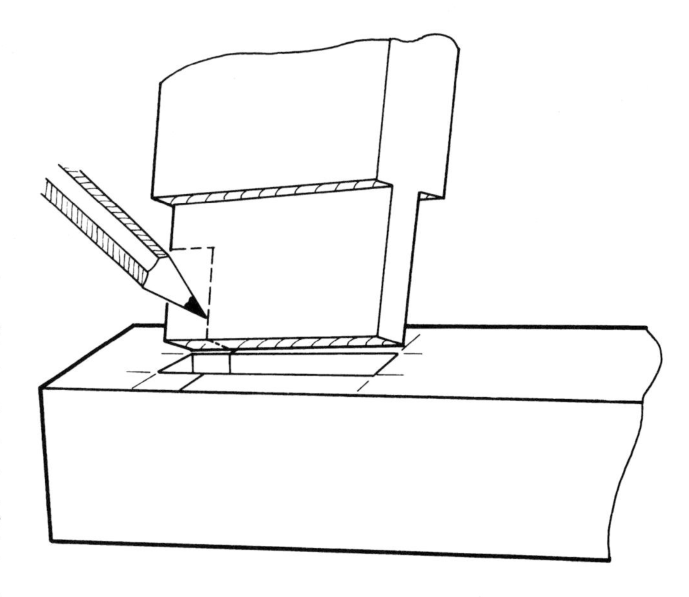

MISCELLANEOUS MORTISE AND TENON JOINTS

There are far too many varieties of mortise and tenon joint to be described in detail, but a few of the more common and potentially useful ones are outlined below. The type of joint deemed to be the most suitable in a given situation has to be weighed and considered against a number of factors, such as the nature of the piece under construction, its purpose and the material from which it is made. Also give some thought to the aesthetic possibilities of the jointing, which can serve a subtle and effective decorative purpose if imaginatively exploited.

STUBBED TENON

This is a simple joint employed in carcass construction and is often used to connect drawer rails to carcass sides. It has its limitations because of the relatively short length of the tenon and the subsequent lack of glued area. It is nonetheless easy and quick to make and has its uses, especially in light pieces.

DOUBLE TENON

This is one of the more common tenon joints used in heavier cabinet framework.

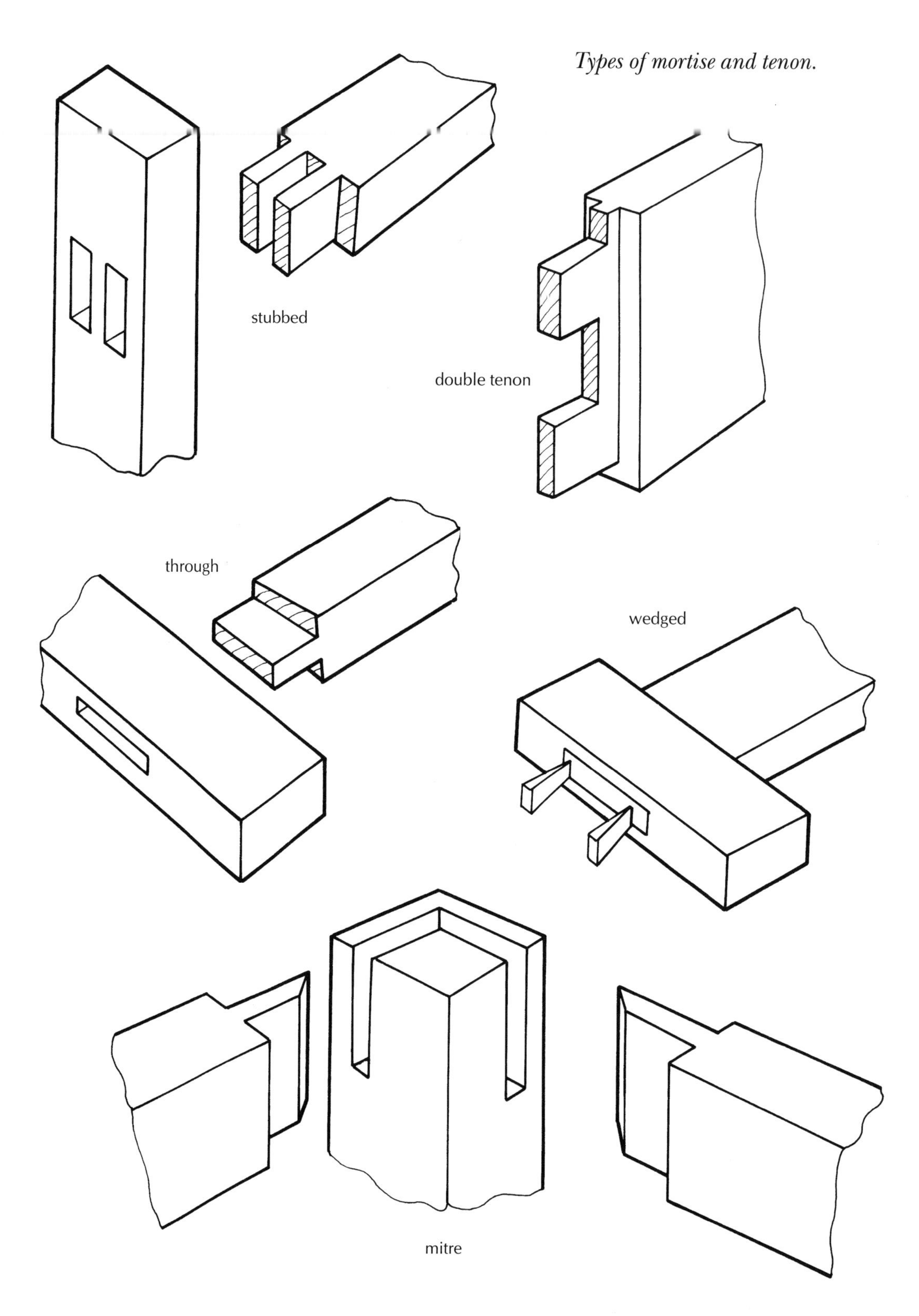

Types of mortise and tenon.

It is similar to its single tenon counterpart and is constructed in the same way. It has the advantage of providing an extremely large gluing area and therefore possesses great strength.

PEGGED TENON

This strong joint comes into its own when the mortise is stopped. It can withstand considerable weight on such items as heavy duty garden furniture. The joint is cut as normal and then temporarily filled with waste timber while a small hole is drilled through the mortise. The waste wood is then removed as a hole, offset at a slightly oblique angle, is drilled through the tenon, which is pushed into the mortise. A tapered pin is then knocked through the joint, pulling the two parts firmly together.

WEDGED TENON

Like the pegged tenon and operating on broadly similar principles, the wedged tenon is quite an intricate joint and therefore only used if the joint is likely to come under extreme pressure. There are two types. For a through mortise two slots are finely cut into the end of the tenon, into which a tapered wedge is driven and glued.

The second type, known as a *fox wedged tenon*, is used for stopped or butt mortises and is slightly trickier to employ. Here the mortise is cut as normal and the two wedges placed into the mortise at the point where they exactly correspond with two slots cut into the oncoming tenon. The tenon is then carefully tapped onto the wedges and glued.

MITRED TENON

This is again a very strong and potentially attractive joint, used when two rails are joined on the same leg or post at right angles. Two 45° mitred joints are cut at the end of each tenon, allowing it to fit neatly into its own individual mortise in conditions where the mortises of standard square tenons would overlap, reduce the available gluing area and deprive the joint of its necessary supportive strength.

CHAPTER **seven**

Dowel Joints

Dowel joints share many of the characteristics of a standard butt or edge joint. They have become increasingly fashionable since the 1970s for a number of reasons. The appeal to industry, which makes extensive use of them, is obvious – they are a cheap, easily produced alternative to the traditional mortise and tenon joint. If sensibly and judiciously applied they can have similar strength to the mortise and tenon yet offer considerable savings in material and time.

A dowel joint consists of a wooden pin fixed and glued into two jointed pieces of timber. So long as the pin protrudes into the socket more than around 35mm (1⅜in), it easily doubles the strength of the pure, unsupported butt joint. It can be used to join the sides of cabinets together or to strengthen housing joints. However, it is most valuable in joining long edges together to form table tops, wall panelling or even stage scenery.

DOWELS

Dowel pins are commonly made of birch and are light and smooth. The dowel can be bought commercially in long, thin strips of varying diameters – typically from 6 to 19mm (¼ to ¾in). Alternatively it can be hand-made, from any desired material and to any specification. It is necessary to purchase or make a cradle and a metal dowel cutter out of steel plate.

Dowel joints have their weaknesses. If carelessly constructed they may weaken and they lack the visual appeal of the traditional mortise and tenon joint for very high

> **TIP**
>
> However they are obtained, it is important that dowels have a groove, cut with a saw or chisel, running up the side. This allows the excess glue and air to escape, without which the reliability of the joint would be seriously compromised. Some dowels are fluted or grooved to increase adhesion.

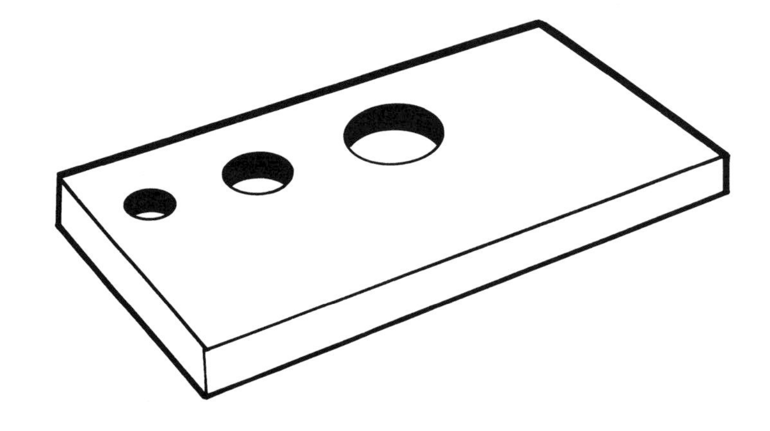

Dowel plate.

quality pieces. However, for many items requiring a strong joint they provide a quick and accessible solution.

CONSTRUCTING A DOWEL JOINT

The first thing to be considered once the decision to fit a dowel joint has been made is what type of dowel should be used and how many dowel pins will be fitted. This will naturally depend on the nature of the edges to be joined and the thickness of the timber. The dowel should be wide enough to provide the necessary strength, but not too wide – if the pin is too close to the surface there is always the risk of splintering. Whenever possible it is advisable to fit two pins onto the joint. Not only will this prevent the joint from coming apart, but it will also stop slippage and movement across the joint.

If two long pieces of timber are being jointed together spacing becomes a consideration. The holes should be marked approximately 175mm (7in) apart. This is only a rough guide and discretion is an important element, but intervals that are too great may not provide sufficient strength, while over caution is costly in time and materials. It is also helpful to use a chamfered or rounded dowel, which will allow easy access to the drilled hole. Having done all this the construction of the joint itself is relatively straightforward:

1. The two surfaces that need to be jointed should be carefully planed and squared in the same way as would be done for any other joint. There is sometimes a tendency to scrimp on this part of the process, with the simplicity of the joint creating a false sense of security – this should be avoided. Surfaces should be planed faultlessly and edges perfectly squared and sawed using a fine-toothed tenon saw. Check that this has been done correctly before proceeding.
2. Next, the two pieces of timber have to be accurately marked out. If the two boards to be jointed are small enough this should present no great problems. Place the two pieces of wood into the workbench vice together. Using a try square and a pencil mark lines across both pieces indicating where the dowel is to be placed. If square sections have to be marked out it may be necessary to use a dowel jig or a template.
3. Using a marking gauge and working from the face side only mark the centre of the dowels. This is best done by drawing a line the length of the two boards.

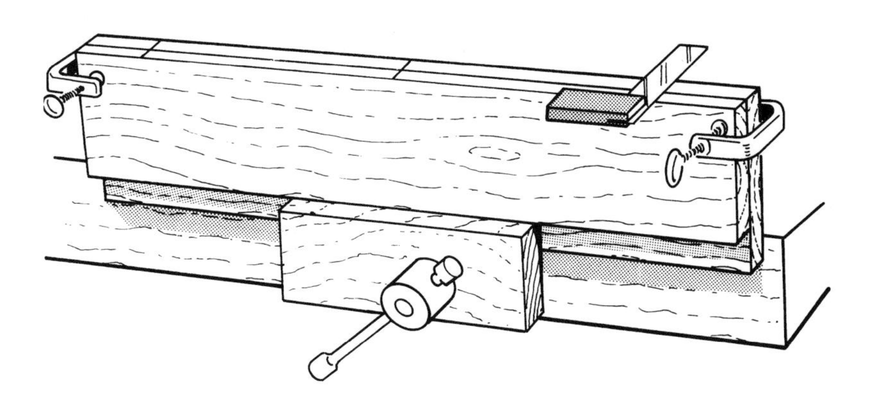

Using a try square to mark dowel positions.

The point at which this line intersects with the squared lines marked previously is to be the centre of the dowel.

4. The holes now need to be drilled. Work only on one board at a time, so remove one piece from the vice, while ensuring the remaining board stays square. Using the appropriate drill bit for the corresponding size of the dowel carefully drill the holes. The biggest problem at this point is to drill the holes vertically. It is very easy to drill at an odd angle, which will create problems later. Drill to a depth of about 25mm (1in). When this has been satisfactorily completed swap the boards over and do the same with the second piece.
5. Cut the dowel to the correct depth. Remember that a very small gap should be permitted at the bottom of each hole for excess glue to accumulate.
6. Sparingly glue the two surfaces of the boards that are to be joined and smooth glue into the holes. This will give the joint additional strength.
7. Push the dowel firmly, but without jarring into one side of prepared holes. If the holes have been drilled accurately this should require only a minimum amount of pressure.
8. Slip the dowel pins into the second series of prepared holes and squeeze the joint together. In some circumstances it may be necessary to cramp the joint gently for a short time.
9. Wipe away any excess glue with a cloth dipped sparingly in warm water.

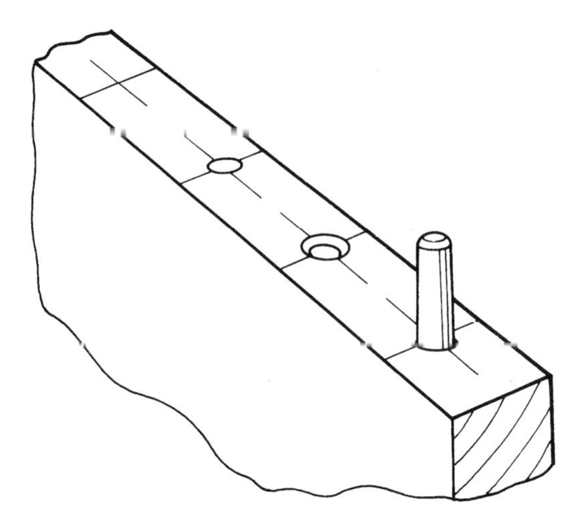

Place the dowel where the lines intersect.

TIP

The size of the drill bit and the dowel may not be exactly the same, especially with commercially purchased doweling. It is sometimes wise to test them out on a piece of waste material before addressing the real job. If the dowel is slightly too broad it can be modified by being knocked through a dowel plate. This can be simply made from a piece of 6mm (¼in) steel bar into which holes on the underside have been bored. If it is home-made ensure that the holes are bored precisely. Dowel plates can also be purchased commercially.

CHAPTER eight

Dovetail Joints

Dovetail joints combine two rare qualities – strength and elegance. Lovingly constructed dovetails add a touch of class to any piece of work and remain one of the few joints that has not been overtaken by the machine. Although there is now machinery available that will cut a neat, precise dovetail, the hand-made article still retains its pre-eminence and for that reason it possesses a nostalgic charm.

One of the reasons why it has held back the march of the machine is that it is one of the more intricate and complex joints to construct. To produce it well demands a high degree of precision and skill, as it allows a much smaller margin of error than its less sophisticated relatives. There is therefore great satisfaction when dovetail joints are mastered and competently constructed. But it would be wrong to talk of the dovetail merely it terms of its traditional attractiveness for it remains one of the strongest of the right-angle joints when the two pieces need joining end to end. This is why it is commonly used in making drawers and in carcass construction.

Its strength emanates from its shape – the dovetail, named after its resemblance to a dove's tail, is little more than a tongue that fits into a corresponding housing socket. The angles of the dovetail and the socket give it its robustness, because when set at right angles, as they would be on a drawer, they are forced together rather than apart, acquiring a mutually shared strength that is not available with other types of joint. While there is no reason why a dovetail joint cannot be glued, it is not a necessity.

Hand cut dovetails fall into three main categories. In the *through dovetail* the joint can be clearly seen on both faces of the corner. In the *lapped dovetail* the joint can be seen on only one face because the other is covered by the lap. In the *secret dovetail* the whole of the dovetail is completely hidden.

To have an understanding of how to construct the dovetail joint it is necessary

The assembled dovetail joint.

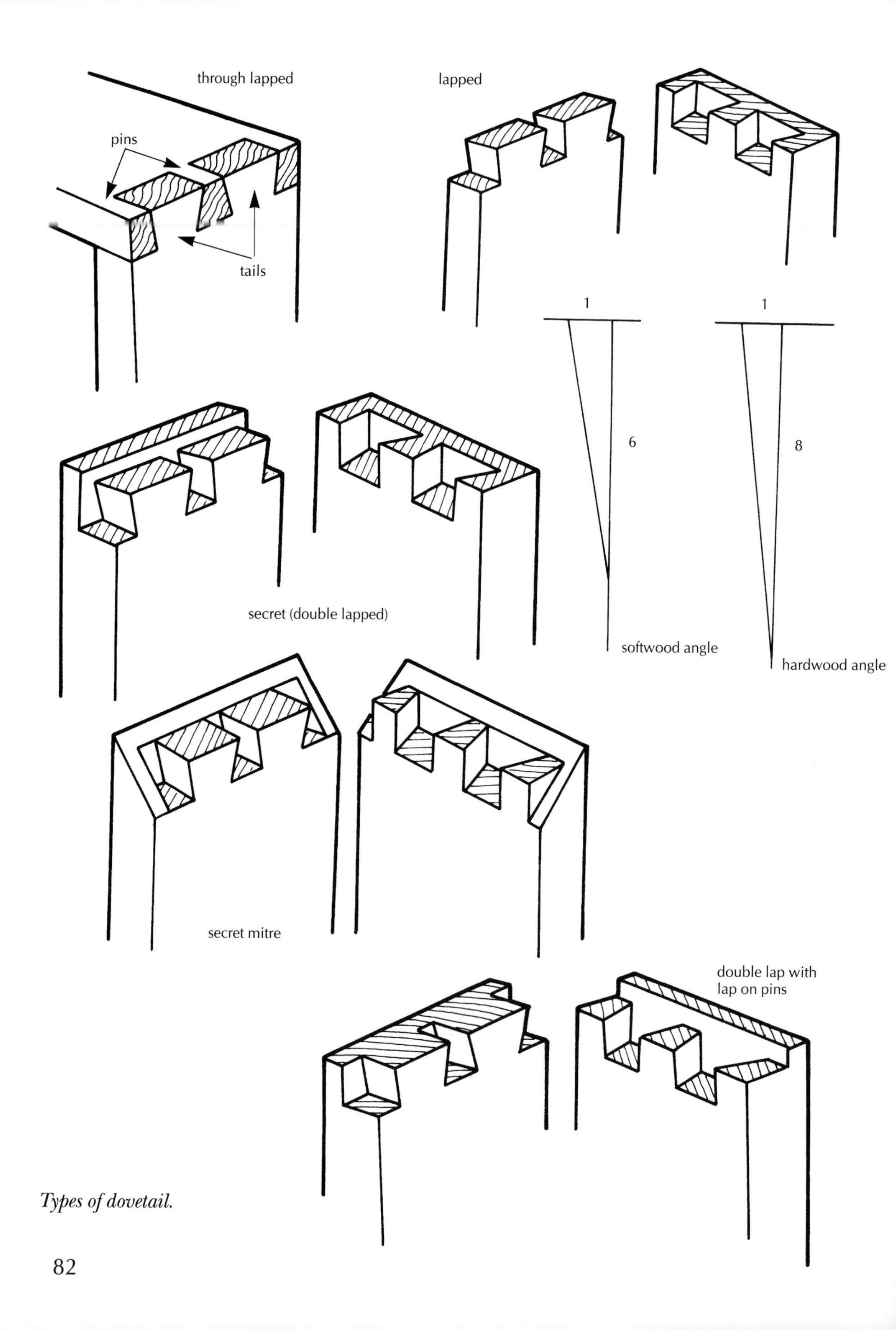

Types of dovetail.

to describe each of these variants separately. Although the underlying principles are similar, there is sufficient divergence to warrant this approach.

THROUGH DOVETAIL

The through dovetail is the strongest of the dovetails and it is also the simplest to make. The bevel or angle of the tails and the size and number of the pins influence the overall strength of the completed joint. A large pin has more strength than a small one, but it is part of the cabinet maker's craft to reconcile the conflict between strength and appearance. A large pin may be strong, but an abundance of end grain on both sides may compromise its comforting appearance.

The angles of the dovetail will vary according to circumstances. A good guide is to employ a slope of 1:6 for softwood and 1:8 for hardwood. Opinions vary as to whether the pins or the tail should be cut first, but it makes little difference on most jobs.

1. The timber components that are to be dovetailed should be planed square and true. A parallel line equal to the thickness of the timber plus 1mm or ⅟16in should then be drawn around the end face. There are differing techniques for achieving this. One method is to use a cutting gauge, but if this tool is used it is imperative that the cut should not penetrate into the wood too deeply as it would then be difficult to remove the cut line from the front faces at a later stage. An alternative method is to mark out using a try square and a pencil. At a later point, mark over the lines with a knife in the places to be cut away.
2. It is then necessary to decide on the size of the pins. It is useful to search out a bevel edged chisel of similar size so that the pin can be adjusted accordingly. An average size for the pin would be around 12mm (½in), but this may of course vary according to the requirements of the piece.
3. Now that the pin size is selected, the two end pins must be marked out equal to the width of the double angle pin. In the example that follows this will be 10mm (⅜in). The end pins must be divided into two equal sizes and their centre lines drawn along the edge of the timber (*see* below). Decide how many tails are required and divide the distance between the end pins into the required equal parts.

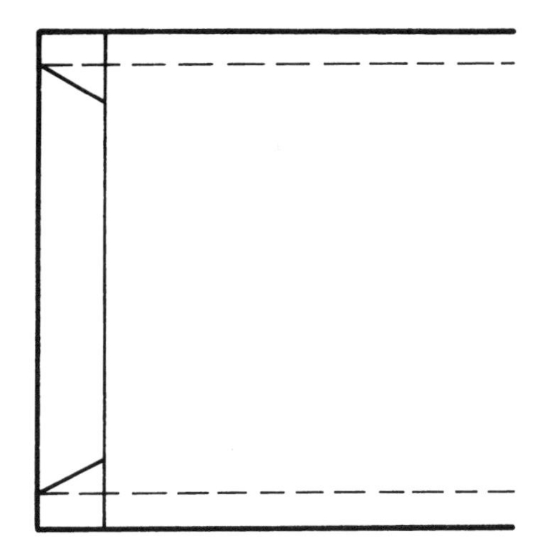

Marking end pins.

4. The line is divided into four equal parts, using an angled rule if necessary. In the example the line is 100mm (4in) long. Mark in a pencil line at points registering 25, 50 and 75mm (1, 2 and 3in). This will establish the centre lines for

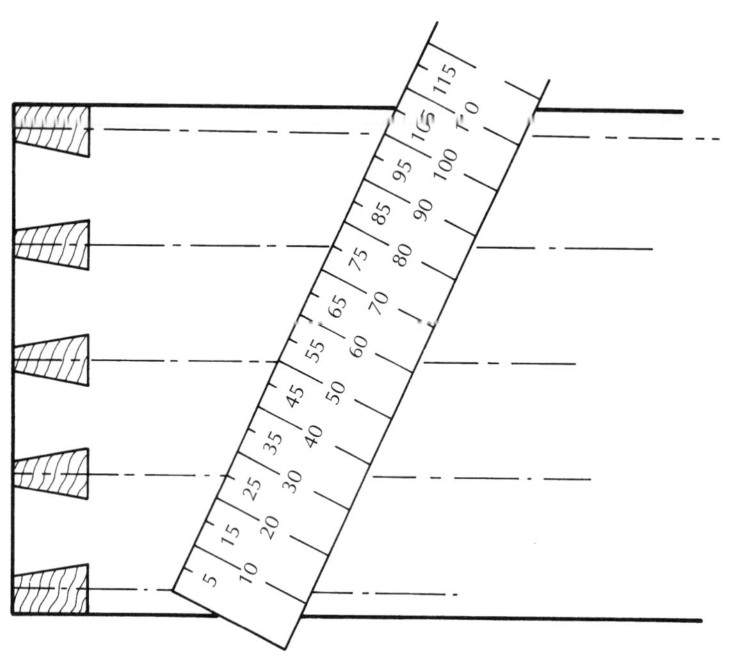

Measuring parts for number of tails.

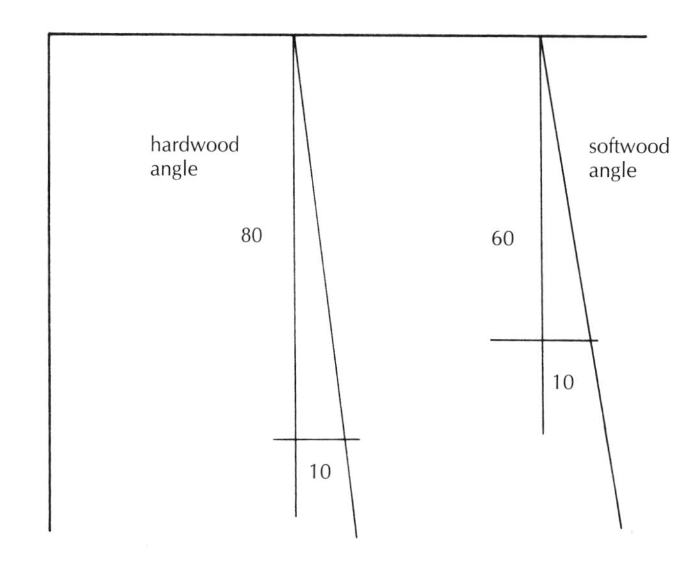

Dovetail angles.

the pins. Mark 5mm (3/16in) either side of the centre lines to produce the double angled pin of 10mm (3/8in).

5. To mark out the dovetail angle use either a dovetail template or an adjustable bevel. To set the dovetail angle for the adjustable bevel, square a line off the face edge of a piece of timber. Measure down 60mm (2⅜in) for softwood and 80mm (3⅛in) for hardwood, then across 10mm (⅜in). Join the two points and set the bevel at the appropriate angle.
6. Place the wood to be cut into the vice. It is helpful to angle the timber so that the dovetails are sawed vertically. Cut all the lines that are sloping to the right first before rearranging the wood in the vice so that the dovetails sloping to the left can be cut. Finally remove the bulk of the waste using a coping saw and finish to the line with a bevel edged chisel.
7. Having marked out and cut the tails it is now necessary to mark out and cut the pins. The best and most accurate way to do this is to use the timber you have cut the tails into as a template.

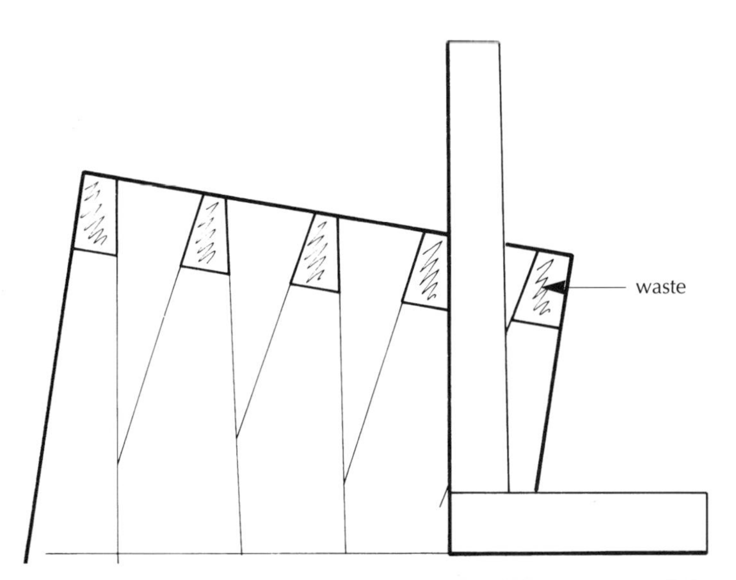

Place the wood in a vice so that you are sawing vertically when cutting the tails.

The smoothing plane is used to help support timber while transferring marks for the pins (below).

However, always make sure that you have numbered your joints because it is extremely easy to make the mistake of transferring them incorrectly.

8. An established way of setting up the timbers to be marked is to place the timber requiring the pin marks into the vice and protruding up to the same

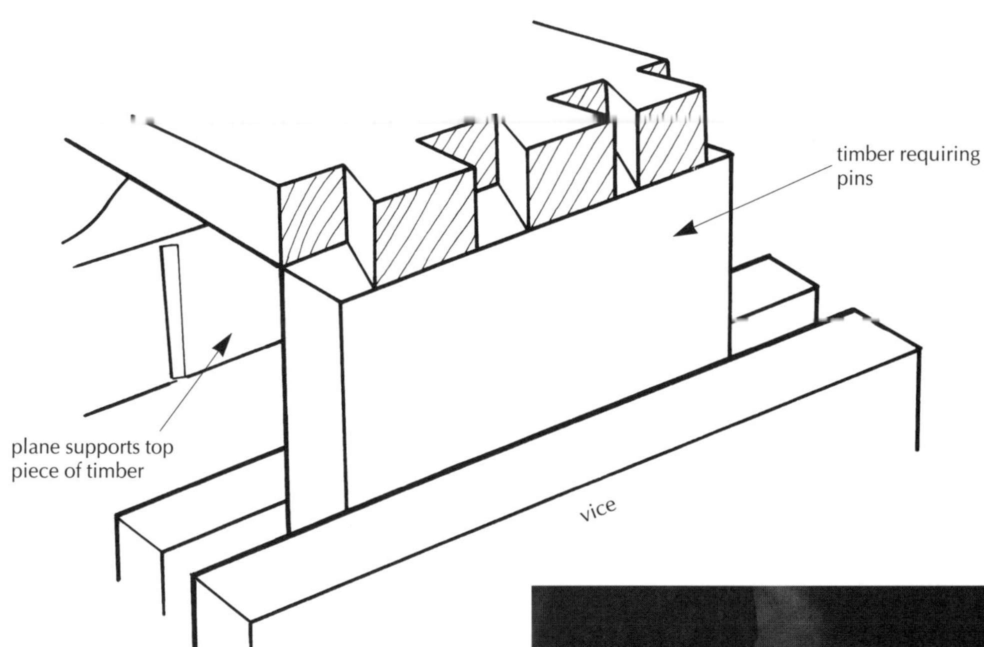

Line up the joint to make sure it is square.

Using a piece of scrap wood and a hammer to assemble the dovetail joint.

TIP

When cutting a dovetail avoid making the mistake of sawing clear of the line with the thought of paring to the line with a chisel later. It is quicker and more accurate to saw straight to the line using a specialist dovetail saw. Mark off the waste wood before cutting. Leave the ends 1mm or 1⁄16in over length to allow for cleaning up afterwards. It is much easier to trim the dovetails to the ground work than it is to trim the ground work to the dovetails.

height as a smoothing plane tipped onto its side. Then place the timber with the tails on top, carefully lining up the joint and making sure that it is exactly square.

9. Making use of a sharp knife, for example a scalpel with a number 11 blade, mark around the tails. Next use a square and a pencil to complete the marking out of the pins. Take great care to saw down the waste side of the line, which should be left just intact. This requires practice, but if done correctly it will ensure good, tight joints and should prevent splitting.
10. Remove all the waste using a coping saw and finish off using a chisel.
11. Some people slightly chamfer the inside edges of the tails to give a steady lead into the pins. This should not be necessary as they should fit together with a few light taps of a hammer, protecting the ground work with a piece of scrap wood.

LAPPED DOVETAIL

The principles behind the construction of the lapped dovetail are the same as with the through dovetail. The main difference is that the dovetail is cut shorter than the thickness of the wood holding the sockets. This type is often used in items where it is necessary for the joint to be hidden, primarily in carcass work and for the fronts of drawers.

The top rail of a carcass is traditionally a lapped dovetail, which can be created in the following way:

1. Mark the amount of lap on the carcass sides.
2. Cut the rail to be dovetailed to the length of the lap lines.
3. Mark out the length of the dovetail.
4. Mark out and cut the dovetail in the same way as with the through dovetail, using a dovetail saw.

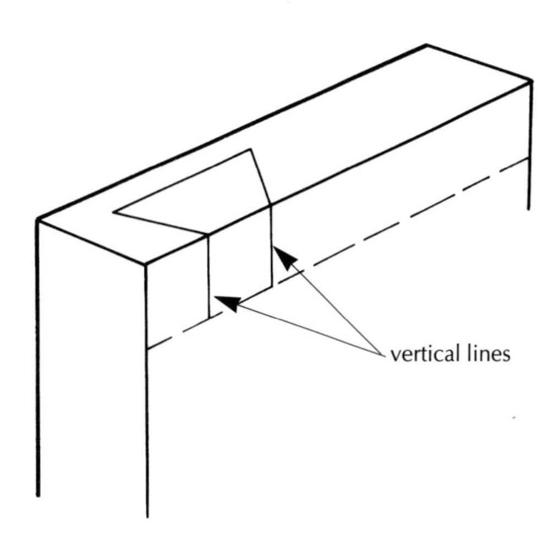

Square the vertical line.

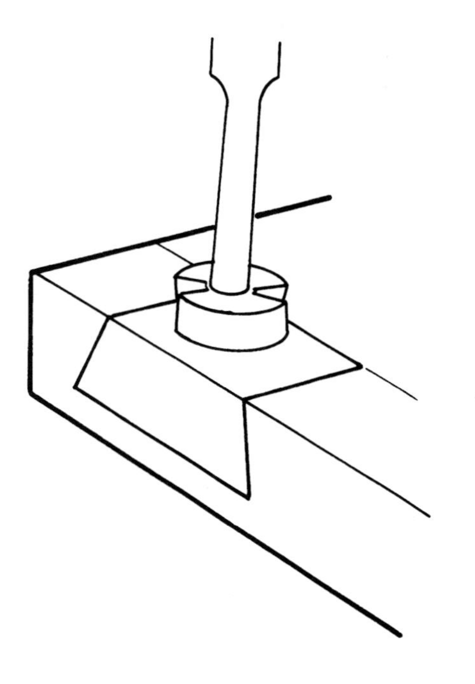

Drill out the bulk of the waste.

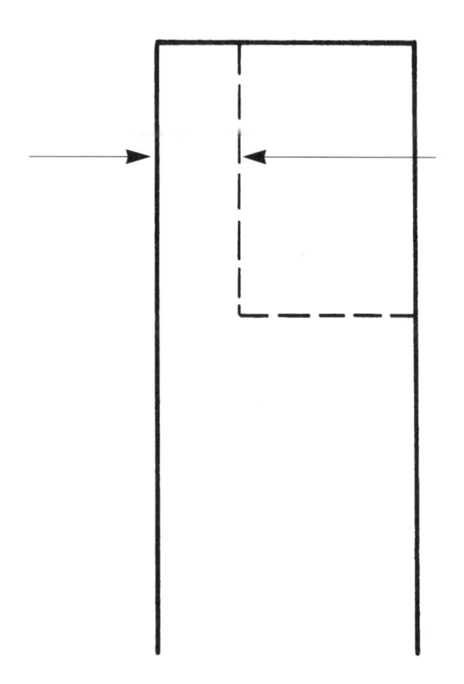

Ensure the lap is parallel.

5. Hold the rail in the correct position using the rail as the template.
6. For the pin mark round with a knife.
7. Square the vertical line from the pin to the correct depth.

To remove the bulk of the waste timber in the sockets use a drill or a router before finishing off and cleaning to the lines with a bevel edged chisel. Now place the rail in the correct position and tap it home using a hammer and a piece of scrap wood to protect the ground work.

A common fault with lapped dovetails is for the front of the tail to come away from the lap. This is normally because the lap is not perfectly parallel. If this fault is identified and corrected the quality of the joint is greatly improved.

SECRET DOVETAILS

These joints (also known as double lapped dovetails) are not as commonly used as the previously described dovetails. They are employed where the joint would blemish a desired decorative effect or where sight of the joint would be intrusive. The pin is made in the usual way, but the projecting lap piece is rebated to create the tail. It is necessary to be careful when creating this joint and it can prove to be a time-consuming affair. A further variation is to create a secret mitred dovetail where not even one end grain is shown. This intricate joint is used only for high quality work.

Paring the waste from between the pins.

CHAPTER 

Veneering

Veneering has been used for thousands of years. It was practised in ancient Egypt and was revived during Restoration times when the over exploitation of oak necessitated the need for alternative methods and materials. During the seventeenth and eighteenth centuries veneers were used with new originality and style for the construction and decoration of furniture. Problems encountered then are even more intense today. Reserves of hardwood are dwindling and so it is no surprise that the use of veneers is being reappraised.

It is unfortunate that in recent years the use of veneers has been associated with poor quality workmanship, when this is in fact far from the truth. The range of veneers and their application will be outlined in an effort to dispel this myth.

CUTTING VENEERS

The manufacture of veneers is a specialized process. It is important that the timber does not contain any defects or its value is reduced. The assessment process needs to discount excessive knotting, ingrown bark or resin pockets.

The timber is softened by steaming and treated to make it ready for cutting. The two most common methods of cutting veneers are rotary cutting and flat cutting or slicing. *Rotary cutting* is used particularly for constructional veneers (hardwood and softwood). It is used a great deal in the production of plywood and laminated work because it is both a cheap and an efficient way of cutting large quantities of veneer. The action of the knife through the annual growth rings causes veneers cut

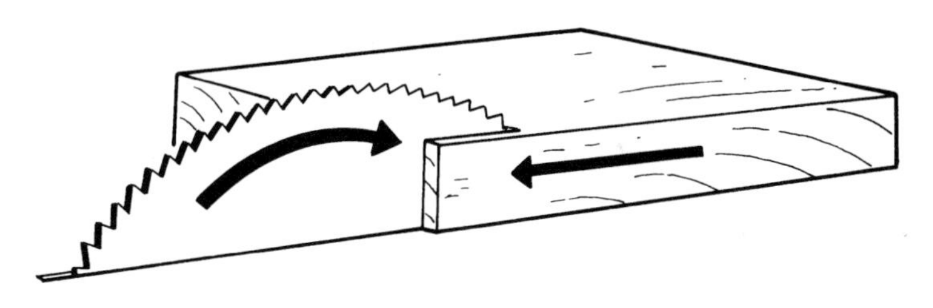

Saw cut veneer.

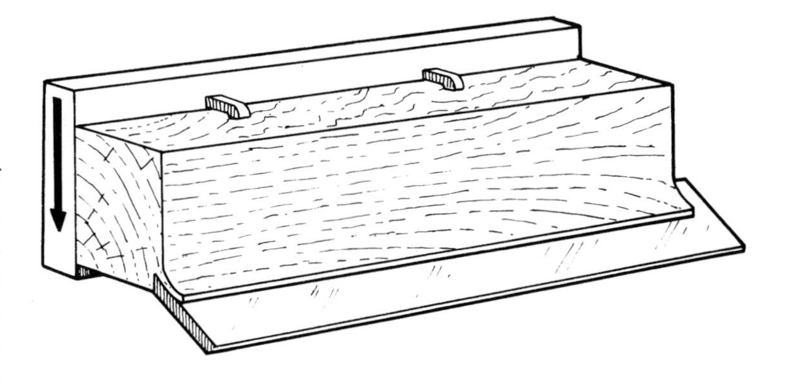

Flat cutting.

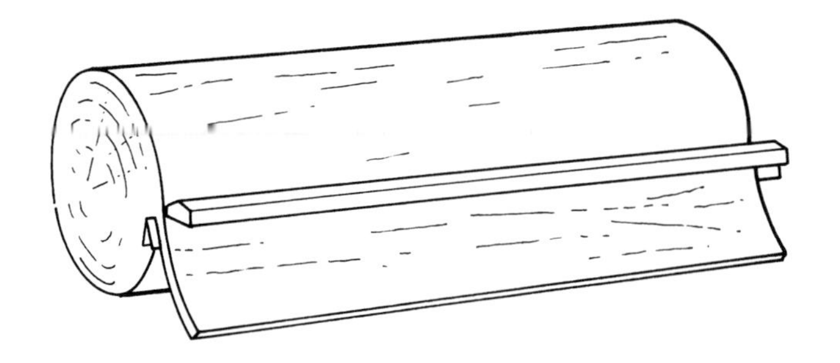

Rotary peeling.

uplifted, serrated frill or feathered figure. This is obtained from the crotch of the tree. *Coloured veneer* is stained artificially and treated with chemical dies to give different colourings.

Ray figured veneer is made from oak. The ray cells within the timber give it a peculiar figure. This is also known as splash veneer. *Striped veneers* are produced by quarter cutting across

in this way to have an exaggerated wavy effect, which is very suitable for decorative veneers, with bird's eye maple being the most common.

Decorative hardwood veneers are usually produced by *flat cutting*. This is a complex process and the shape of the figure produced depends on the nature of the wood used and the way the timber is cut. Flat slicing produces a smooth cut, the best results being obtained when the cut is at right angles to the grain.

In *reconstituted veneers* the veneer is reformed into a more desirable pattern. It is sometimes stuck end to end to create effects like waves or herringbone. *Constructional veneers* are mostly straight grained, being used in laminated work for curved slopes in furniture construction. These veneers are over 1.2mm (1/16in) thick.

Crown cut veneers are from flat slices cut at a tangent to the grain of the wood. This gives rise to a bold oval formation in the centre of the sheet and stripes at the edges. Crown cut veneers are obtained from ash, sapele and others. The sycamore and other wavy grained woods give *curly figured veneers,* which show alternating dark and light bands. 'Fiddleback' sycamore veneer is so called because it is used for the backs of violins.

Curl veneer has an attractive figure when cut at right angles to the grain to give an

the growth rings. In *butt veneer* a distorted grain causes a complex figured veneer when back cut on a rotary cutter. This is obtained from the butt of a tree trunk on species such as ash.

Burr veneer is expensive – it contains a compact pattern of rings and dots. It is found in maple and used for intricate furniture work. Burrs are defects on the trunks of trees. They are formed by fungi or insect attacks and make the wood very difficult to plane.

DECORATIVE VENEERING

The imaginative use of veneers can create artistic and pleasing effects. Some of the most commonly used techniques are listed below.

In veneering in the form of *stringing* (or lines) a fine strip of wood – such as boxwood, ebony or dyed wood – is laid into the surface. This separates areas of veneer by creating light or dark lines.

Banding is done with decorative side grain or cross-grain sections of coloured wood about 1mm (1/16in) square. *Cross banding* has narrow cross-grain edged borders that may be found on table tops, the border of which may be edged by stringing.

For *bookmatching* place one veneer on its face, turn the following veneer over to produce a bookmatch. Care has to be taken to ensure consistent grain configurations. In *random match* leaves are not matched for grain and need not be of the same width. *Quartered* veneer involves cutting four consecutive leaves diagonally and then matching them at the side and butt to produce the diamond quarter.

One means of producing both simple geometric patterns and sophisticated pictures is to use the technique of *marquetry*. An effect of cross-graining is produced by cutting patterns from several different sheets of veneers, inlaying them and then laying them on a single sheet used as a foundation.

Bookmatching.

BUYING AND STORING VENEERS

Veneers are usually kept in groups of four to allow matching. It is important that the sheets are placed in sequence. Veneers are priced according to their size and quality. A degree of wastage will inevitably occur and this needs to be taken into account when calculating the amount required.

Veneers need to be stored flat between polythene or newspaper, away from light due to the risk of fading. It is also advisable

Random match.

Quartered veneer.

to have an even temperature and atmosphere. After a time veneers may become fragile, so it is best not to handle them until it is necessary. If veneers are intended for decoration they must be numbered with chalk in the order in which they were cut. The ends need to be taped to prevent splitting.

VENEERING TECHNIQUES

Traditional hand veneering uses hot animal glue. A warm working environment is essential because the glue loses its adhesive properties as it cools and gels. Adjacent leaves of veneers will have been cut consecutively from the same log and will therefore be a good match. The leaves should be numbered with a pencil or chalk to maintain this match.

Using medium-density fibreboard, lightly sand the ground work using a coarse grade of abrasive paper to score the surface and produce a key for the hot glue. For solid timber, work a toothing plane across the grain. It is possible to make a simple toothing tool by hammering several nails or panel pins through a thin piece of wood, so that they just protrude to produce a comb like effect on one side. Speed is essential as the glue cools and gels very quickly, so have all your materials ready at hand.

1. Mark the centre of the veneer joint on the ground work on each side, with a straight edge.

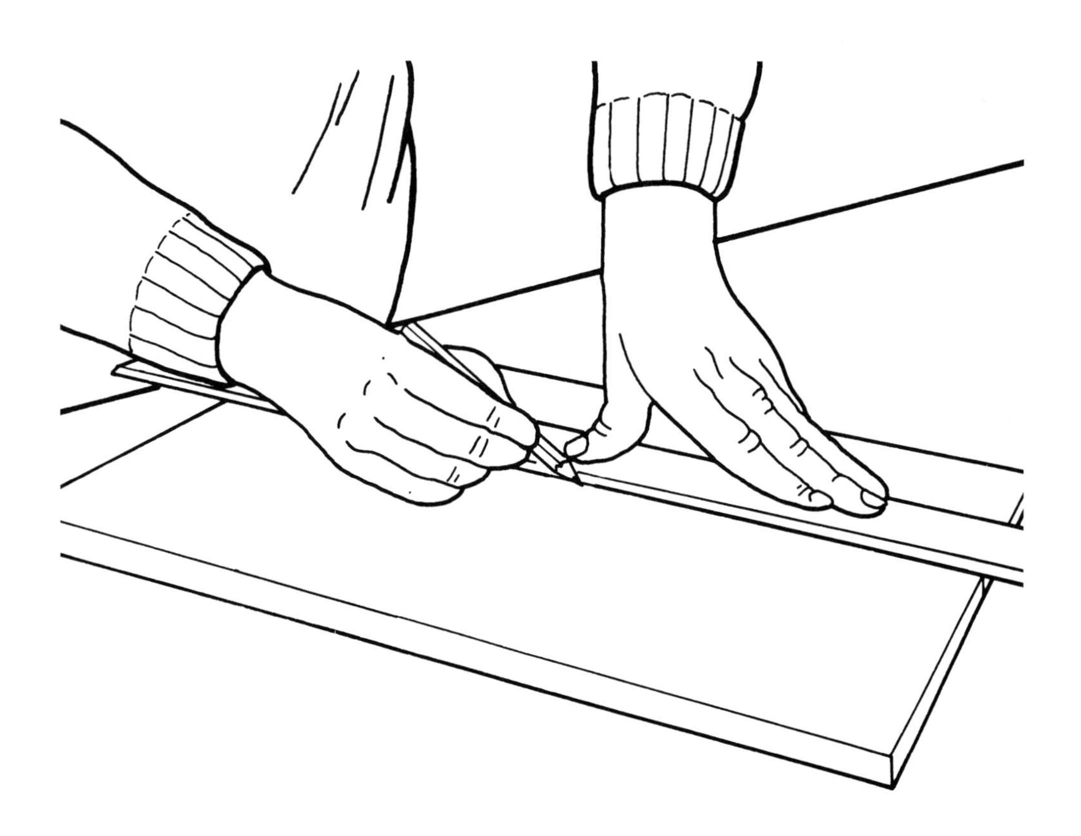

2. Warm the brass strip of the veneer hammer to avoid chilling the glue – this can be done by laying it on a warm electric iron. Using a brush, apply hot glue thinly and evenly to the first side of the ground work.
3. Place the veneer on the ground work allowing approximately 12mm (½in) overlap over the centre line of the join, then press the veneer into place with the veneer hammer.

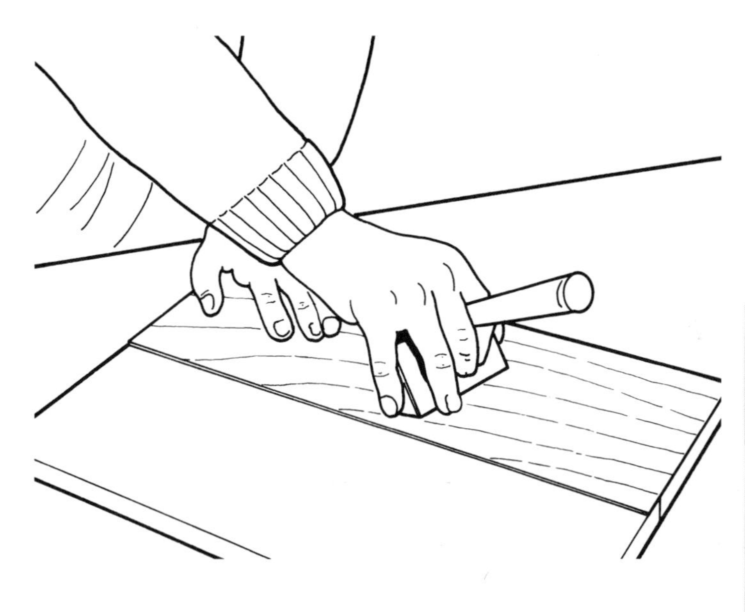

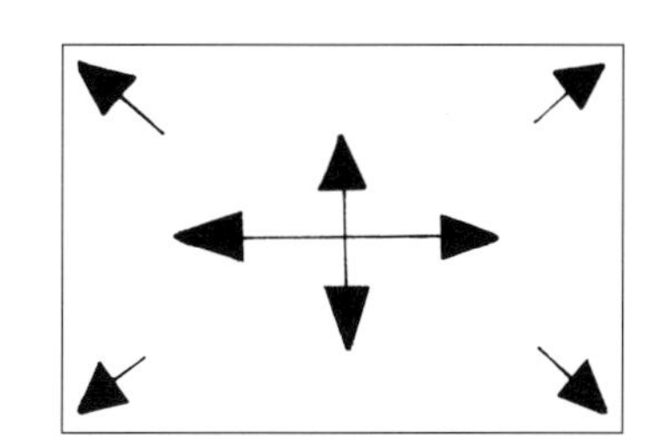

4. Work out the excess glue and trapped air bubbles to the edges, with a firm pressure.

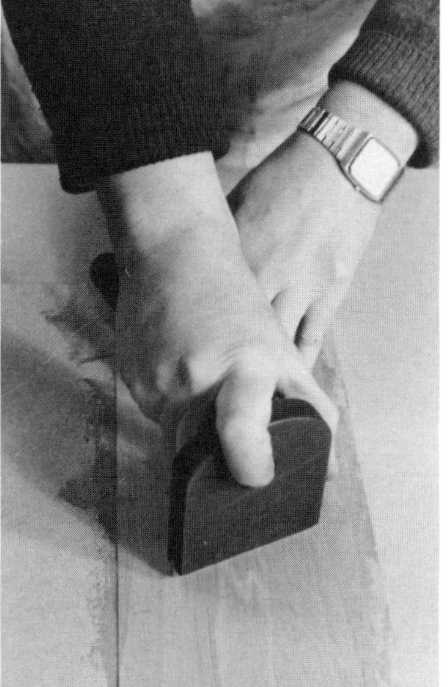

Work out excess glue using firm pressure.

5. Work from the centre of the veneered panel outwards.
6. Apply the glue to the other side of the ground work as before, place the next leaf of veneer in position, again allowing 12mm (½in) overlap over the centre line of the joint.

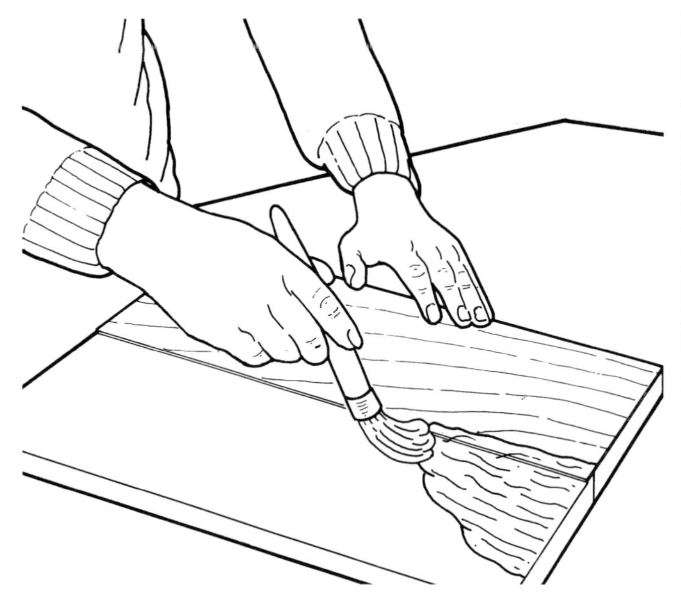

Apply glue to the other side of the groundwork (above).

7. Work out the surplus glue and air with the veneer hammer.

8. Position a straight edge on the centre line of the joint and with a scalpel cut cleanly through the two veneers.

Removing the top veneer.

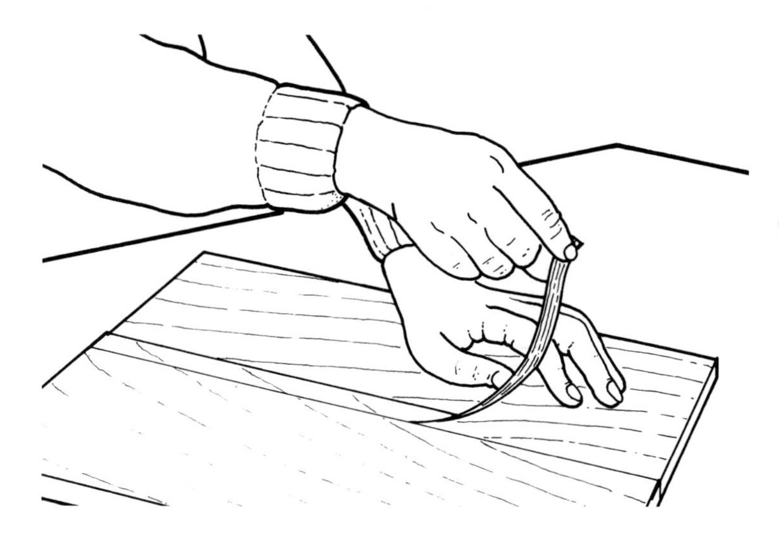

9. The top veneer should peel off easily.

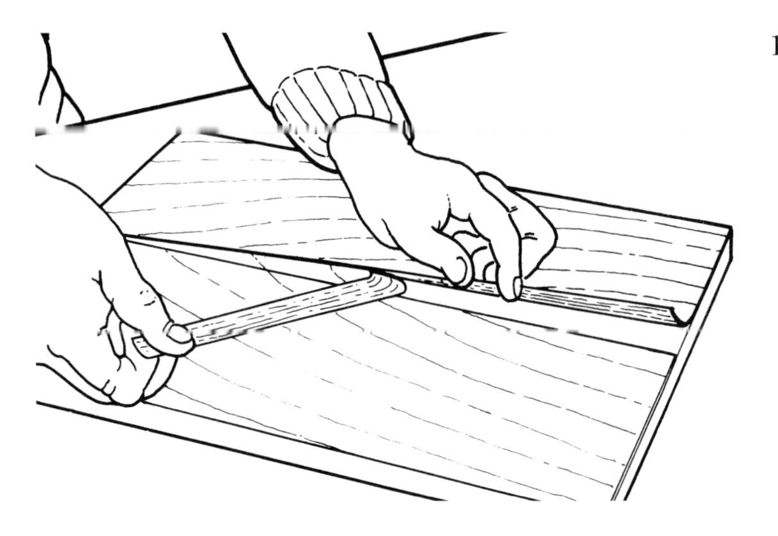

10. To remove the lower veneer, gently peel back the top layer and lift out surplus veneer. If required, run more glue along the edge of the joint and force down using the veneer hammer. Remember that the glue is hot and wet and if you use too much the pressure on the joint may cause the veneers to stretch, resulting in a small overlap that will need cutting away again.

Removing the lower veneer.

11. Press down the joint with the veneer hammer.
12. Wipe away any surplus glue with a damp cloth and tape the joint to prevent shrinkage, which will result in unsightly gaps.
13. If the glue gels before the veneer has been laid satisfactorily, reheat it by placing a damp cloth on the affected area and running a warm iron over the surface. Press the veneer down using the veneer hammer.

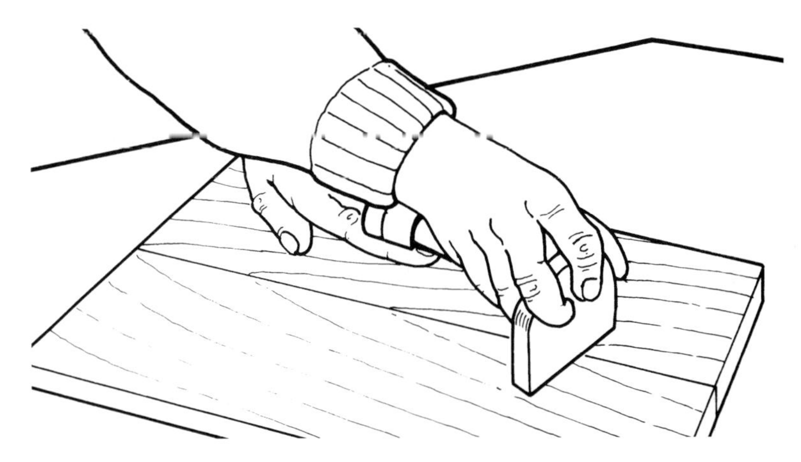

Press the joint down with the veneer hammer (right).

Tape the joint to prevent shrinkage of the veneers.

> **TIP**
>
> As a general rule, what you do to one side of the ground work you should do to the other. If the opposite side will not be seen, a cheaper veneer of a similar thickness may be used to balance the board and stop it warping.

PRESS VENEERING

With this method the veneers are jointed together before being laid. The veneers are cut together with a straight edge, the edges placed together and jointed with glued tape before being placed into position on a glued ground.

The balancing veneer is fixed at the same time. The prepared ground is then placed between two blocks or thick sheets of wood, with a sheet of newspaper or thin card protecting the face of the veneers, and pressure is applied. The veneer should always be cut slightly too large, in case it slides.

There are a number of different types of press used for this application, ranging from a small screw down press – similar to a traditional stationery press – for small work areas to large screw down, hydraulic or vacuum presses. Pressure can also be applied using cross bearers and G-cramps. The technique is as follows:

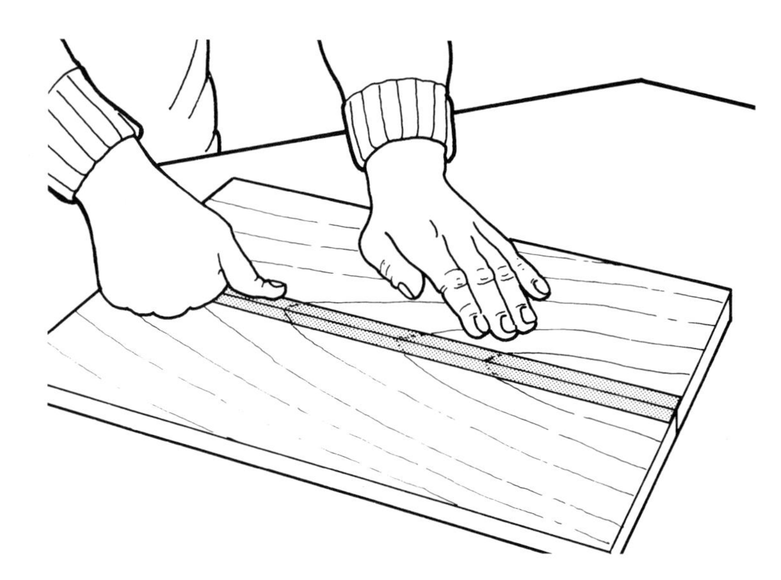

1. Cut a straight edge on the veneers to be jointed (*above*).

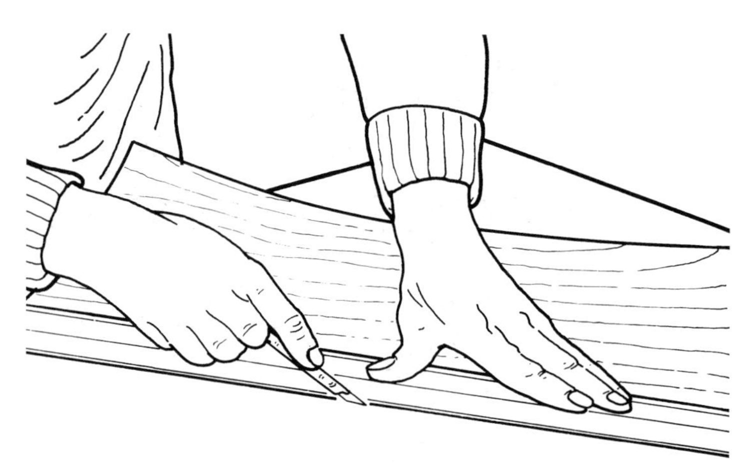

2. Position them together.

3. Pinch the veneer together and tape across the joint.

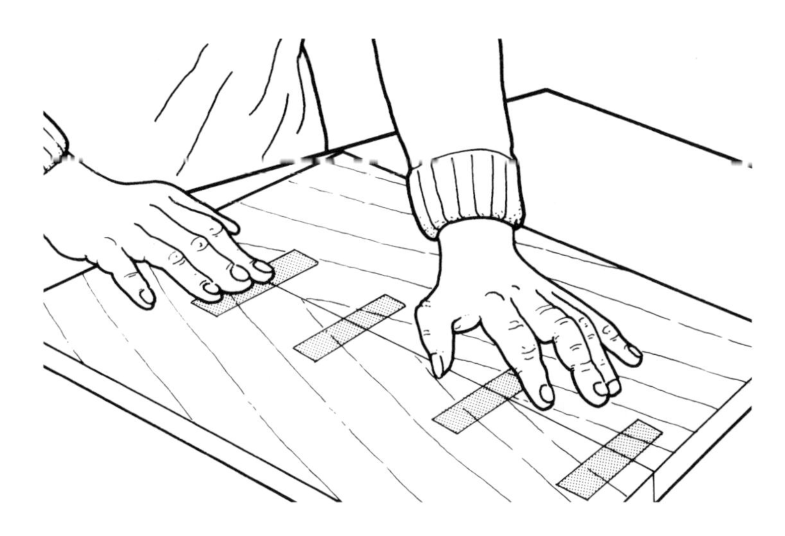

4. Tape down the whole length of the joint. Lightly sand or score the board to be used.

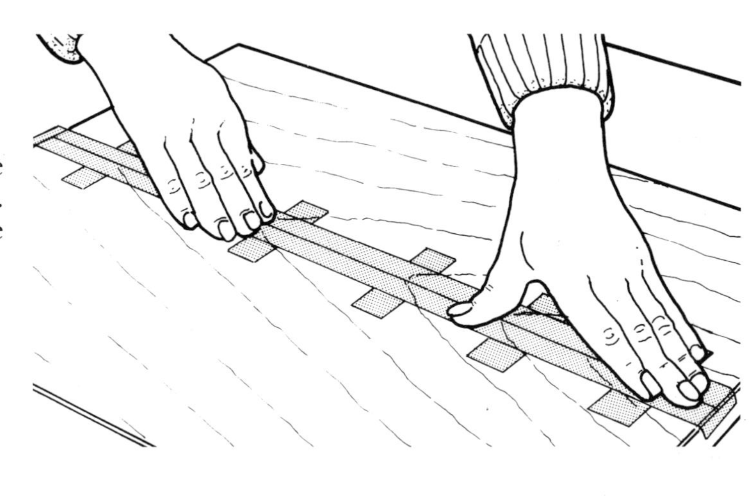

5. Put an even coat of polyvinyl acetate adhesive on the board. A wallpaper seam roller is useful for this task.

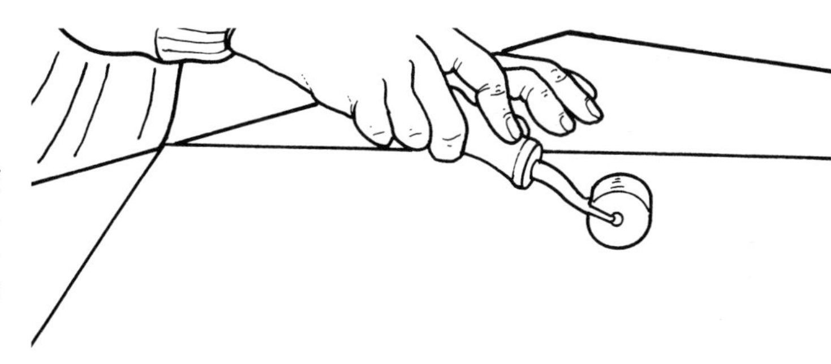

6. Position the veneer on the board and smooth it down by hand.

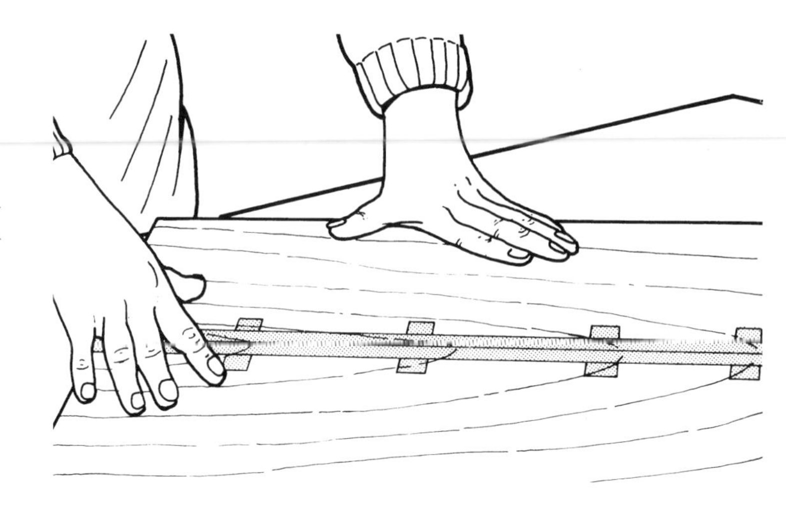

7. Put the piece in the press and cover it with a sheet of lining paper or newspaper. Apply pressure to bring the platens together.

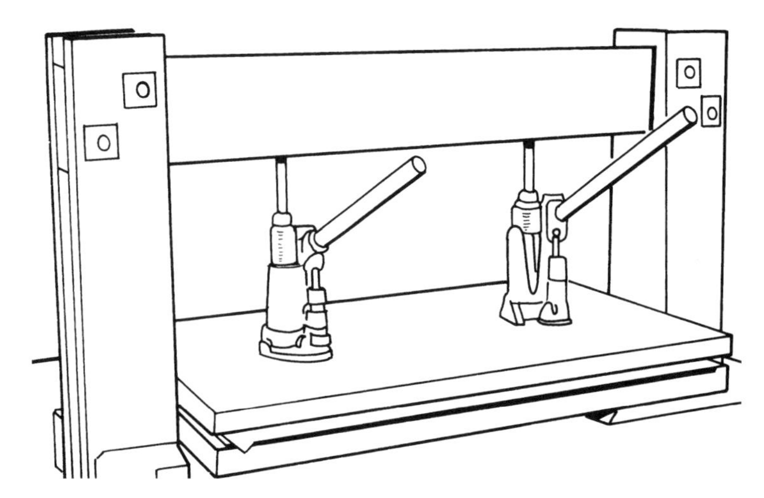

SANDBAG VENEERING

This is a traditional way of veneering curved shapes. It is first necessary to construct a rough wooden box. Put the ground work onto a plywood sheet approximately 25mm (1in) thick. Place the frame around the ground work. Now fill it with sand.

Pack it tightly and tamp it down until it fits the ground work. The box should then be covered with a board. Turn the box over. Remove the ground work, and the veneers tailored to suit the adhesive applied and then put into the sand former. Place the plywood sheet on the top and use weights to apply the pressure.

VACUUM VENEERING

This is used to veneer curved work and to produce laminated shapes around a male former. A plastic or rubber bag and a vacuum pump will be needed – the bags come in sizes from 1200 × 1200mm (47 × 47in) to 3000 × 1500mm (118 × 59in). Place the ground work on a drilled and grooved board to facilitate air extraction and apply the vacuum pump. Modern breathable fabrics are sometimes used, but tend to be less effective than the traditional materials.

CHAPTER ten

Carcasses

The moment of truth comes when you collect the accumulated sections of work and attempt to put them together. This is a potentially terrifying moment, particularly in the construction of a large carcass. There are two underlying principles that have to be adhered to: the carcass must be strong and it must be square. Any joint connecting a front, back or side panel that has not been made accurately will distort the whole carcass. Drawers may not fit snugly, doors may swing open and internal joints may collapse under the strain. If one then takes into account the accumulated wear and tear over time caused by constant usage, the ravages of temperature and weather and the vagaries of uneven floors and walls, then it is essential that the skilled cabinet maker does all that can be done to limit the potential for deterioration. The remainder of this chapter will concern itself with just that, through a description of some of the ways in which carcasses can be constructed strongly enough to withstand all that is thrown at them.

Constructing the carcass perfectly square is the best defence against distortion, but other measures will additionally strengthen it. This can be achieved through the prudent selection of the materials that form an integral part of the workpiece or by the addition of appendages specifically designed to reinforce vulnerable sections.

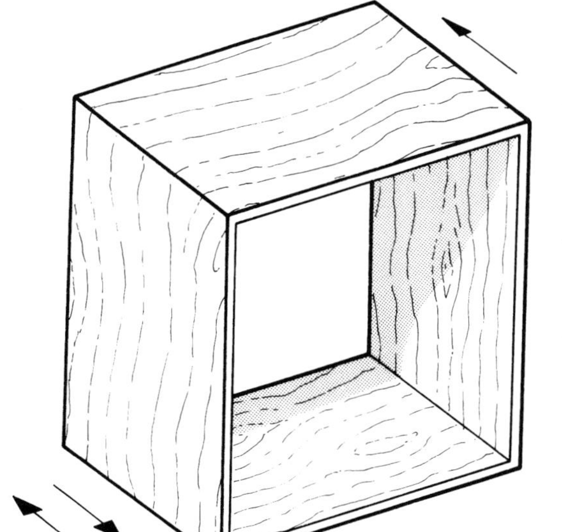

Grain directions.

SOLID WOOD CONSTRUCTION

When preparing a carcass for construction one of the primary considerations should be the degree of shrinkage to be expected from the timber that is being used. Solid wood and plywood should not be thoughtlessly combined as the amount of shrinkage that accompanies these different timbers is widely divergent. That is not to say that they cannot be used in a complementary way, it is simply necessary to bear the shrinkage in mind at the planning stage.

Solid wood construction on its own is rarely undertaken these days at a large scale commercial level – when it is used, it is often for specially commissioned pieces. It is important to make sure that the direction of the grain of the timber selected

TIP

There are a number of ways to increase the overall strength of a carcass:

- Applying side pilasters.
- Applying a central pilaster.

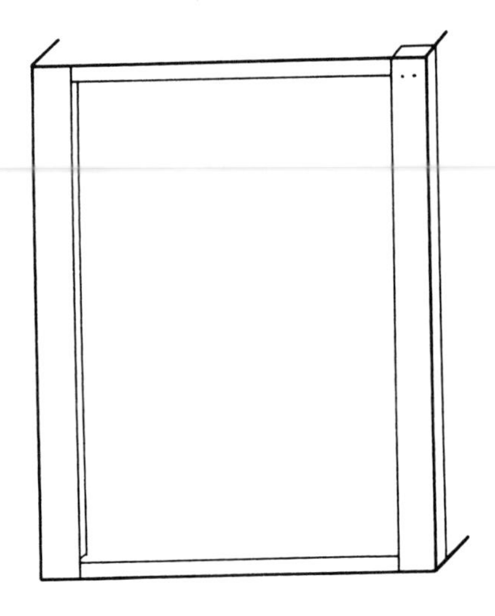

Side pilasters.

Central pilaster.

A rigid apron.

H frame.

- The construction of a rigid apron.
- Using an H-frame.
- Making use of fixing shelves.
- Having a strong carcass back.
- Fitting a strong plinth.

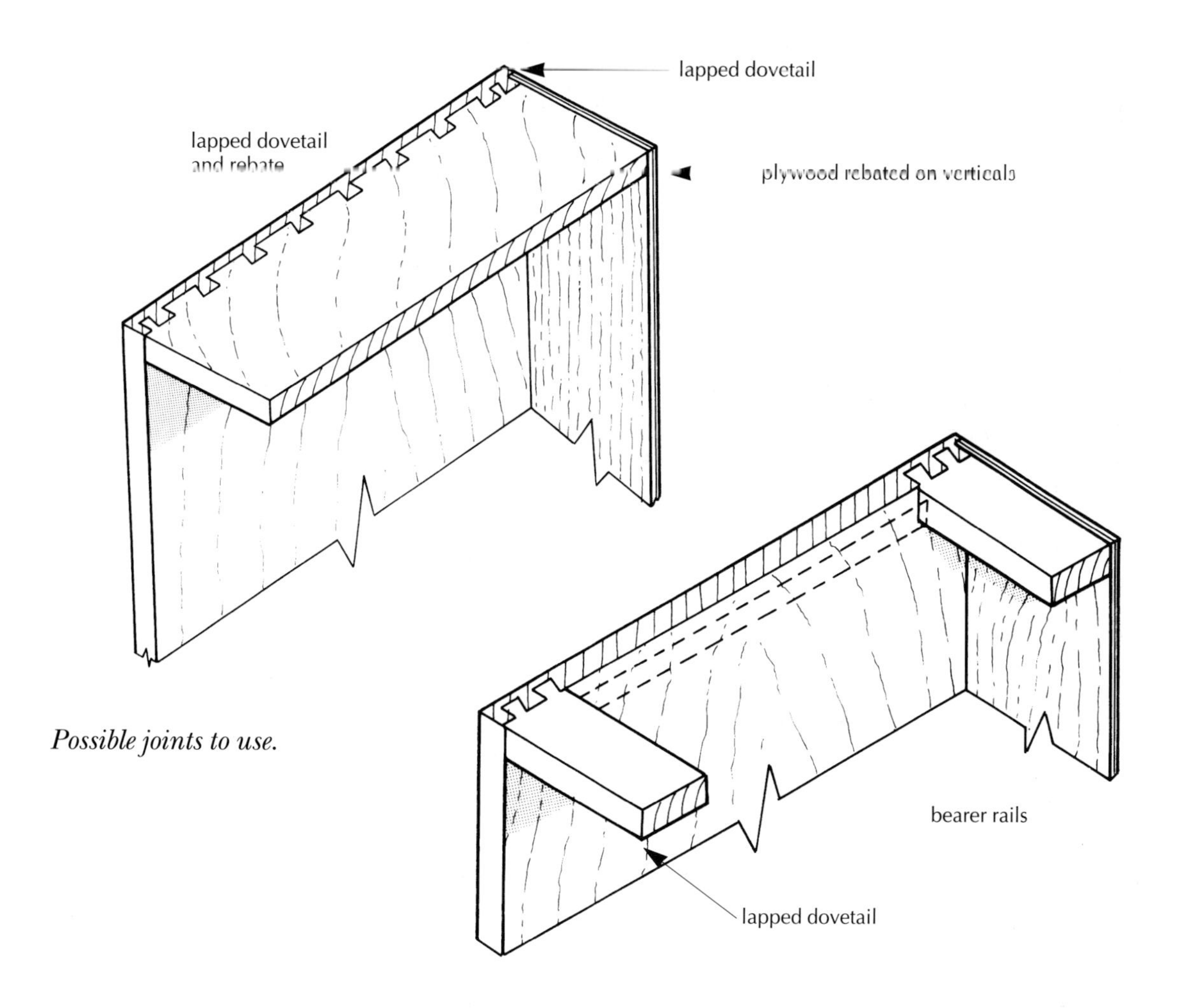

Possible joints to use.

allows for shrinkage uniformly across the width of the boards.

To hold this type of construction squarely and securely in place the judicious use of various types of dovetail joint is necessary. Alternatively, the dowel joint is an option, as is the housing joint. The selection of joint employed on the piece is a matter of personal preference and will be influenced by the time available to complete the task, the function and purpose of the individual piece and the cost. This is particularly pertinent for solid wood construction as these jobs tend to be commissioned and the requirements of the customer have to be considered carefully.

CARCASS BACKS

Carcass backs have three main jobs to perform:

- To close the opening of a piece of furniture.
- To provide the carcass with additional stiffness and rigidity.
- To increase the amount of weight at the back of the piece to prevent it from falling over forwards.

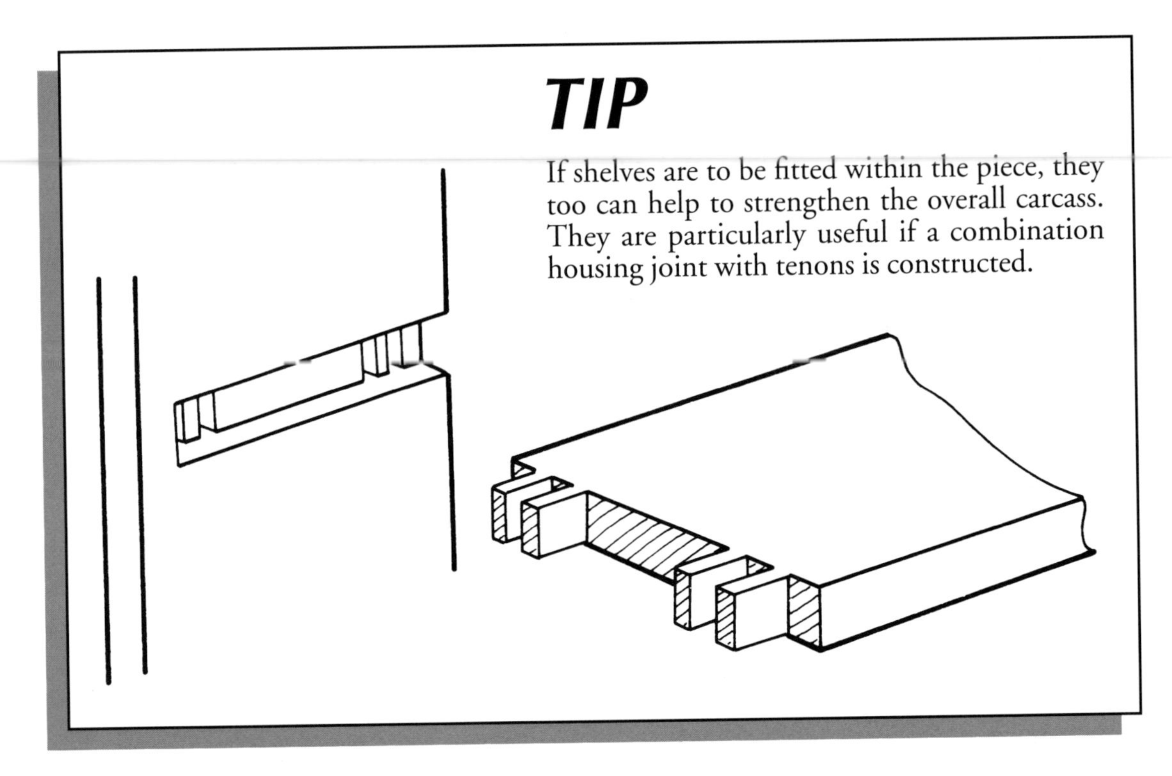

TIP

If shelves are to be fitted within the piece, they too can help to strengthen the overall carcass. They are particularly useful if a combination housing joint with tenons is constructed.

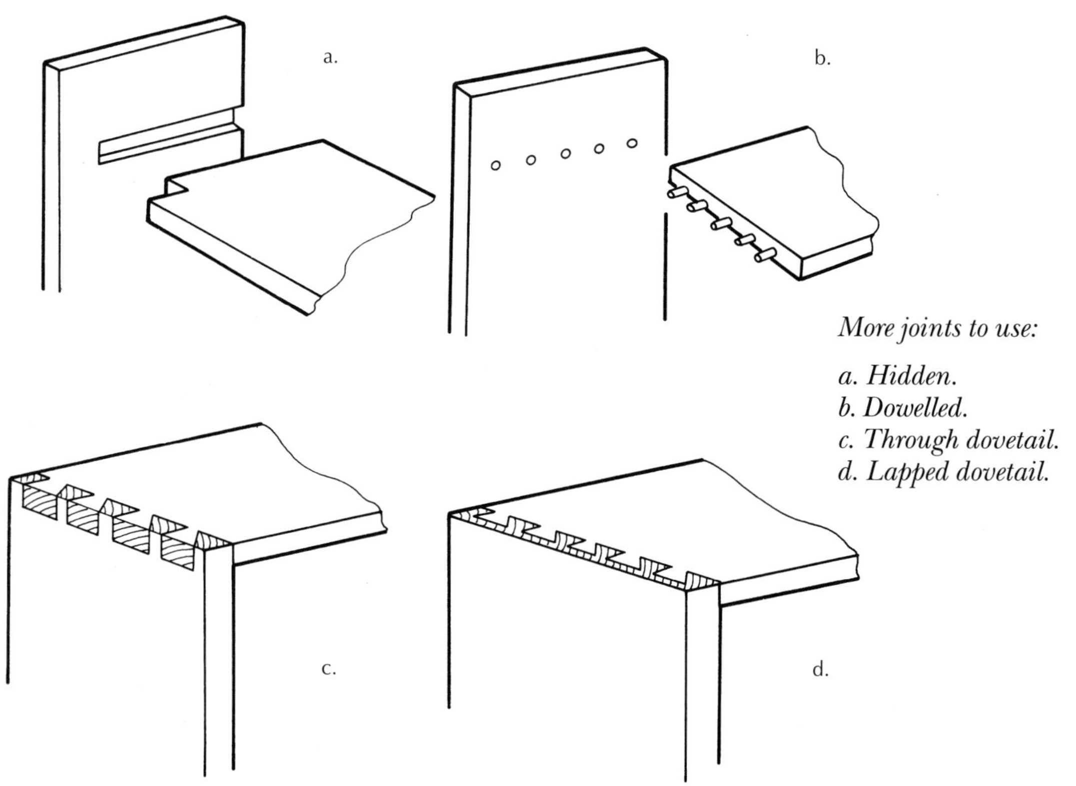

More joints to use:

a. Hidden.
b. Dowelled.
c. Through dovetail.
d. Lapped dovetail.

The last of these is more important on tall items where the centre of gravity is higher than usual, and on items that stand up against a wall. The most common example would be a wardrobe: the weight of the doors opening can tip the wardrobe over. The thickness of the back board will determine to a degree the available weight for the construction of the doors. A heavy back board for a little more leeway.

There is much to be said for using a traditional panel construction for its pleasing symmetrical appearance. It also has the advantage of being able to pull the carcass square. The backs can be made of a range of materials such as laminboard or a stiff plywood. A good though expensive choice would be solid Lebanon cedar as this will provide the necessary weight and possesses the unusual aromatic quality of being repellent to moths. (Anyone who knows the distinctive pungent smell of moth balls will find this side effect more than a little appealing.) For items that are of a poorer quality, hardboard and thin plywood are sometimes used, but this is not recommended for higher grade work as it lacks the strength and weight of the alternative.

CABINET BACKS

Traditionally the best cabinets were backed with solid timber, even though there is not quite the emphasis on weight that there is on taller items. It is now customary to construct the back in the form of frames or panels. Their thickness largely depends on the material from which they are made – around 9mm (⅜in) in solid wood or 6mm (¼in) in plywood. The stiles and rails range from a width of 50 to 75mm (2 to 3in) and are approximately 16mm (⅝in) thick.

Jointing varies tremendously and as with most decisions has to be made in the light of each circumstance. Mortise and tenon joints are customary for framing, but doweled joints are just as acceptable nowadays. Carcass backs may be made of natural or synthetic board, either veneered or plain.

Panelled cabinet back.

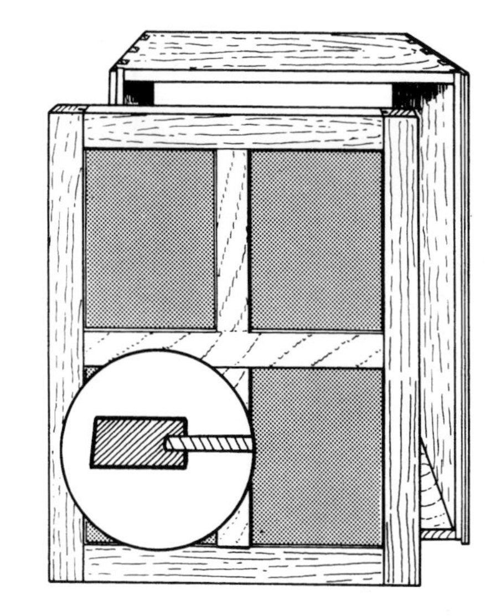

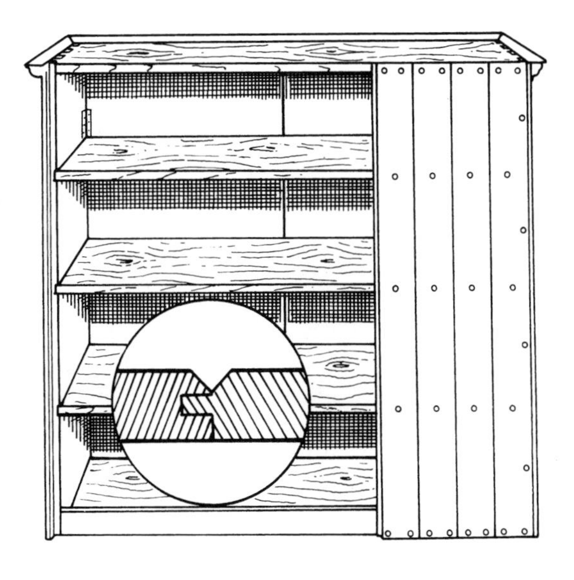

Tongue and groove boards on a cabinet back.

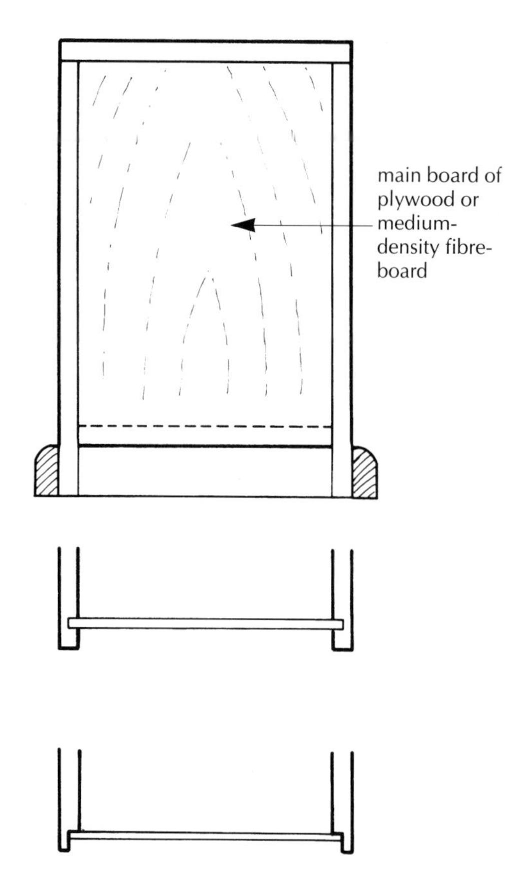

Board fitted into a groove or rebate in a cabinet back.

PLINTHS

On work undertaken in the traditional manner plinths are never recessed, but are made to protrude forwards and outwards. Modern designs have tended to shift away from this for practical as well as aesthetic reasons. The advantage of the recessed plinth is that if it is damaged by shoes, a vacuum cleaner or children's toys, the damage is below the shadow line and therefore out of sight. On quality furniture the plinth is made independently of the rest of the piece. When the final product is fitted together it is fixed with glue blocks, dowels or buttons.

CORNICES

The cornice is the top moulding on a piece of furniture. Cornices came into their own in the eighteenth century when the classical influence was very much to the fore in all areas of the arts. They faded out of design for much of the twentieth century, being considered too elaborate in an essentially utilitarian age. However, the cornice now seems to be having something of a renaissance and is again becoming a fashionable feature in furniture design.

Cornices are generally moulded and then applied to a frieze board. They can be applied directly onto the carcass or can be designed as a separate unit that is then dropped onto the top of the carcass and held in position with glue blocks or dowels.

SHRINKAGE BUTTONS AND SHRINKAGE PLATES

Wood is a naturally porous material. It will absorb or expel moisture at varying rates, depending upon the nature of the wood and the atmospheric conditions. For this reason it is important that any timber waiting to be used needs to be carefully stored and well looked after. Once the timber

Making shrinkage buttons.

a. Dimensions for the button.
b. Countersink screw holes.
c. & d. Make several at a time.
e. Fix with a groove.
f. Typical button positions on a table top.

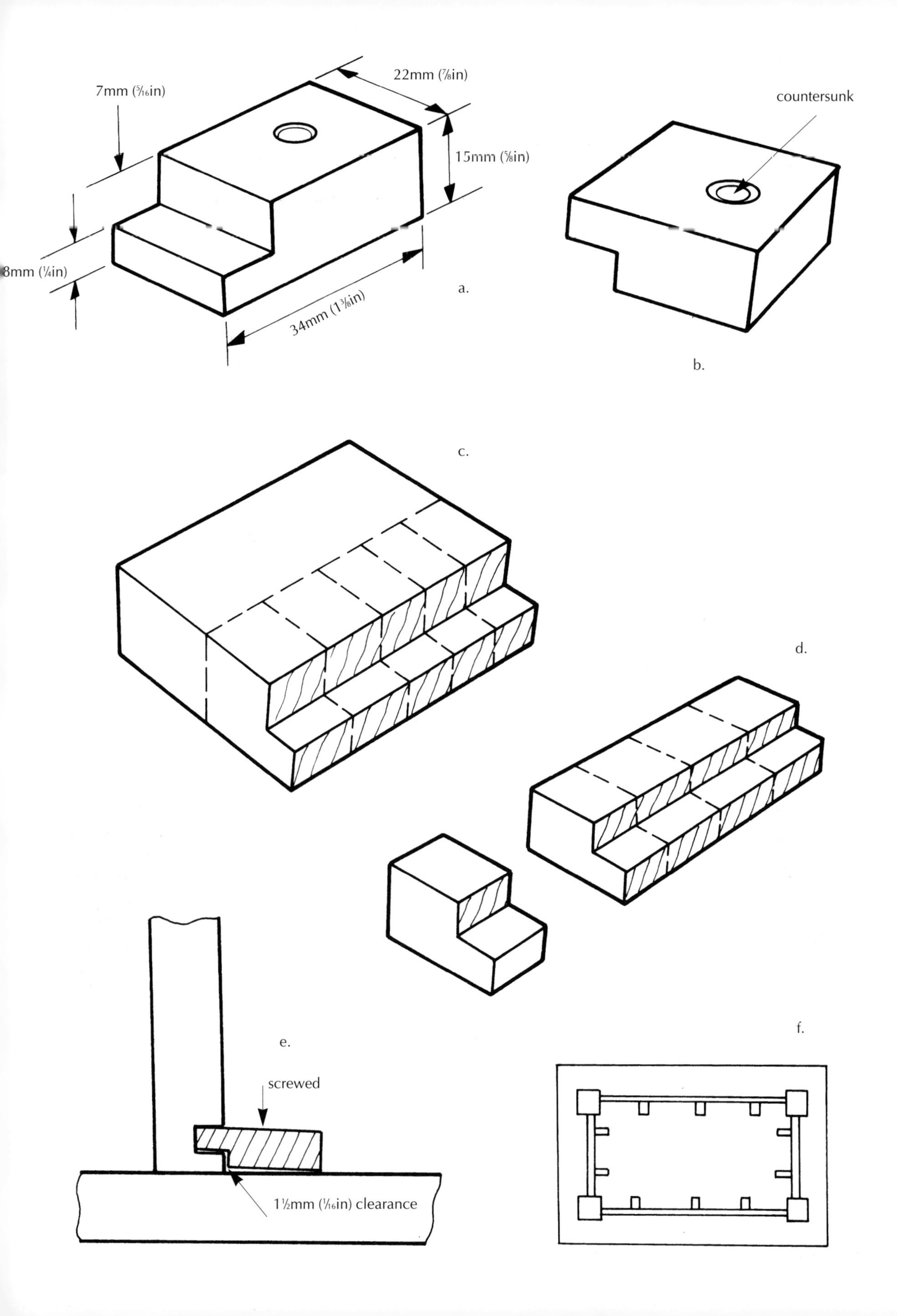

7mm (5⁄16in)
22mm (7⁄8in)
15mm (5⁄8in)
8mm (1⁄4in)
34mm (1 3⁄8in)
a.
countersunk
b.
c.
d.
e.
screwed
1½mm (1⁄16in) clearance
f.

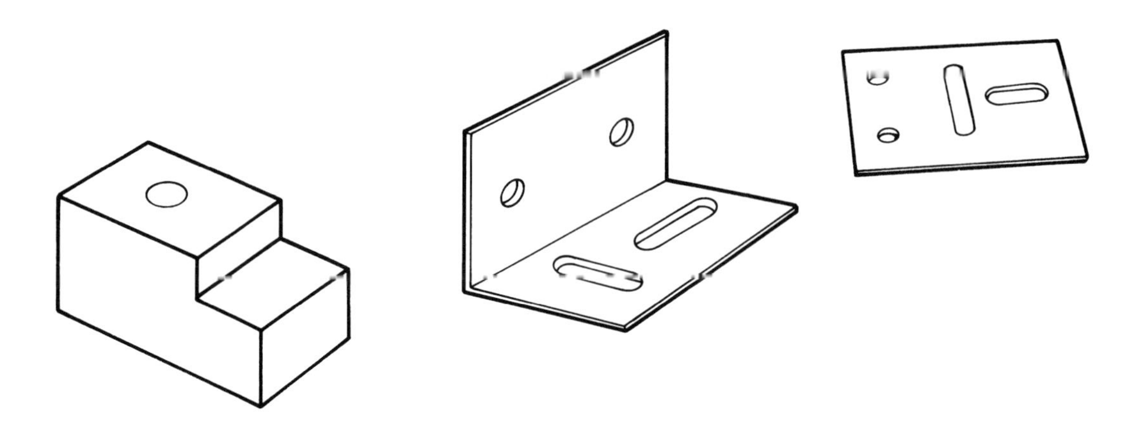

*Button (*left*) and metal shrinkage plates.*

forms part of a piece of furniture this propensity to absorb or expel moisture does not go away. However, its effects can be limited by sealing the timber with a good quality polish. When polishing a piece, remember not to be over-zealous as some allowance needs to be made for movement (generally across the width of the timber). Therefore, when fixing a solid top onto a table or chest of drawers, attaching a plinth or applying a cornice, allowances need to be made to compensate for this movement. Failure to do so may result in the top splitting or bending.

The solution to this problem is reasonably straightforward and involves the use of shrinkage buttons or shrinkage plates. To make *shrinkage buttons* choose a piece of timber of an appropriate size. Rebate the timber across its end so that time can be saved by making a number of buttons simultaneously. Cut the buttons to the required size. The buttons are used for screwing down such things as tops to solid wood carcasses and so their exact size needs to be individually established. When this has been done cut a slot about 1.5mm (1/16in) above the height of the tongue of the button onto the receiving timber. This is so that the button will pinch when it is later screwed into place. The buttons can then be screwed down into the normal position for fixing the table top to the frame.

It is also possible to use metal *shrinkage plates* in a similar way. These can be purchased commercially in either an angle form or in a straightforward flat form and are screwed into position. It is important that the screws are placed to allow the top to slide freely across the width.

CHAPTER eleven

Putting it all Together

It is useful to think of the frame of a piece of furniture as the skeleton onto which the main body of the piece rests. The frame is usually visible, so it is the subject of much decoration, though in some types of cabinet it may be hidden by sheeting. The frame may be constructed to support a table top, an elevated cabinet or a seat for a chair or stool.

FRAME CONSTRUCTION

Most frames operate on common principles and there is little leeway to deviate from them. Perhaps the best way of illustrating this is to demonstrate how the frame for a simple table is made; this can then be applied to other items such as cabinet stands and chairs with relatively little modification.

TABLE-TOP SUPPORT

A standard frame used to support a table top consists of four vertical legs or supports held in position by front and back rails (side rails) and two end rails. These are connected and held firm by mortise and tenon joints or dowel joints. This same principle holds regardless of the number of legs, with only slight modifica-

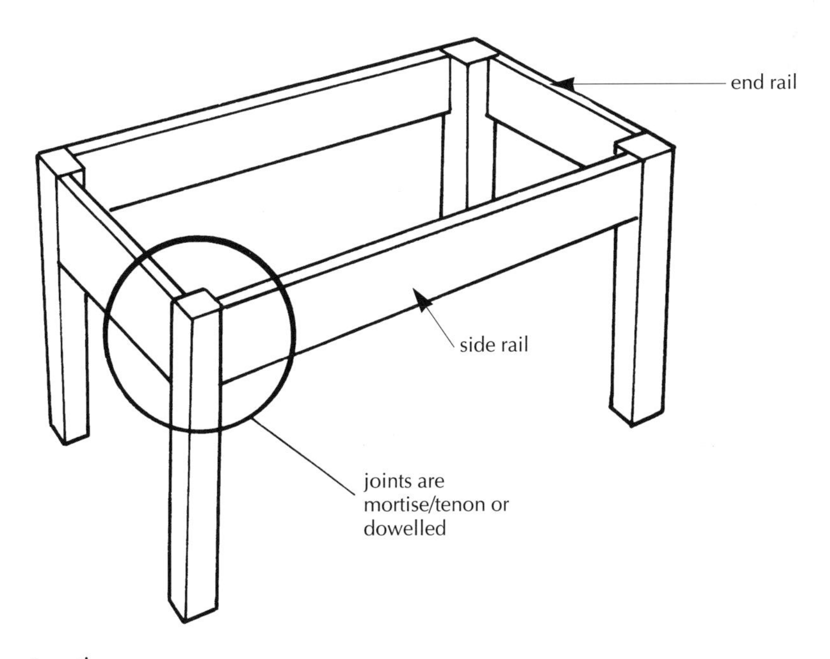

Framed table construction.

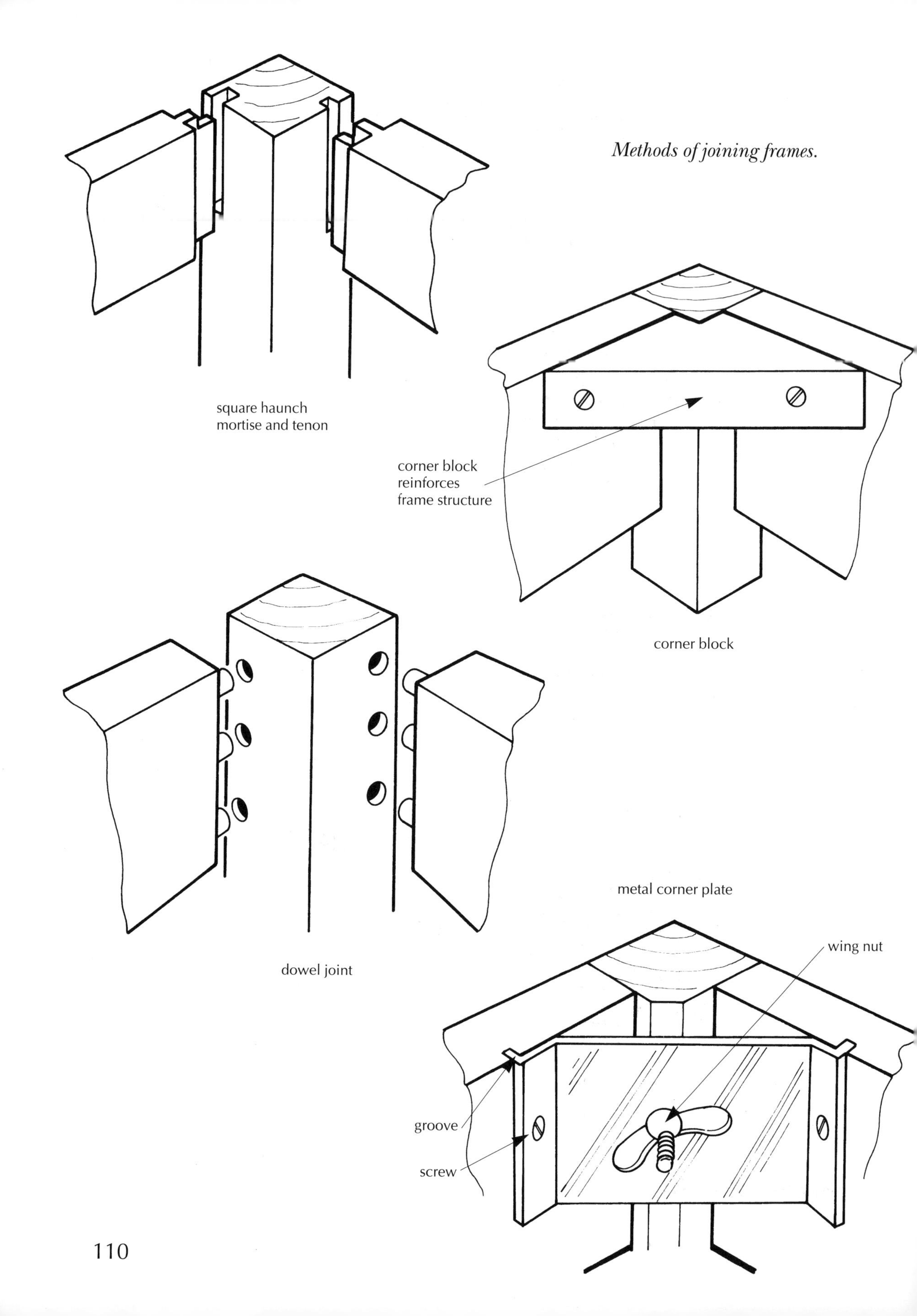
Methods of joining frames.
square haunch
mortise and tenon
corner block
reinforces
frame structure
corner block
dowel joint
metal corner plate
wing nut
groove
screw

tions to the angles of the rails and joints needing to be made in some instances. The rigidity of the piece is strongly influenced by the depth of the rails in proportion to the height of the table: the deeper the rails, the less scope there is for lateral movement. While this is not a problem for cabinet stands it presents a practical difficulty in tables where an overly deep rail makes it difficult for people to sit down. This difficulty can be overcome in particularly tall pieces of furniture by the addition of bottom stretchers or rails.

Stretchers are helpful in reinforcing the triangulation of the structure out of all proportion to their relatively delicate appearance. They are normally mortise and tenon jointed and can be laid out in a H-frame, set across the perimeter of the structure mirroring the top rail or arranged in an × formation from corner to corner, the rail having the joint in the middle.

Corner blocks can be used to strengthen the structure further and to tie the rails together. For a temporary structure such as a card table, corner blocks can be replaced

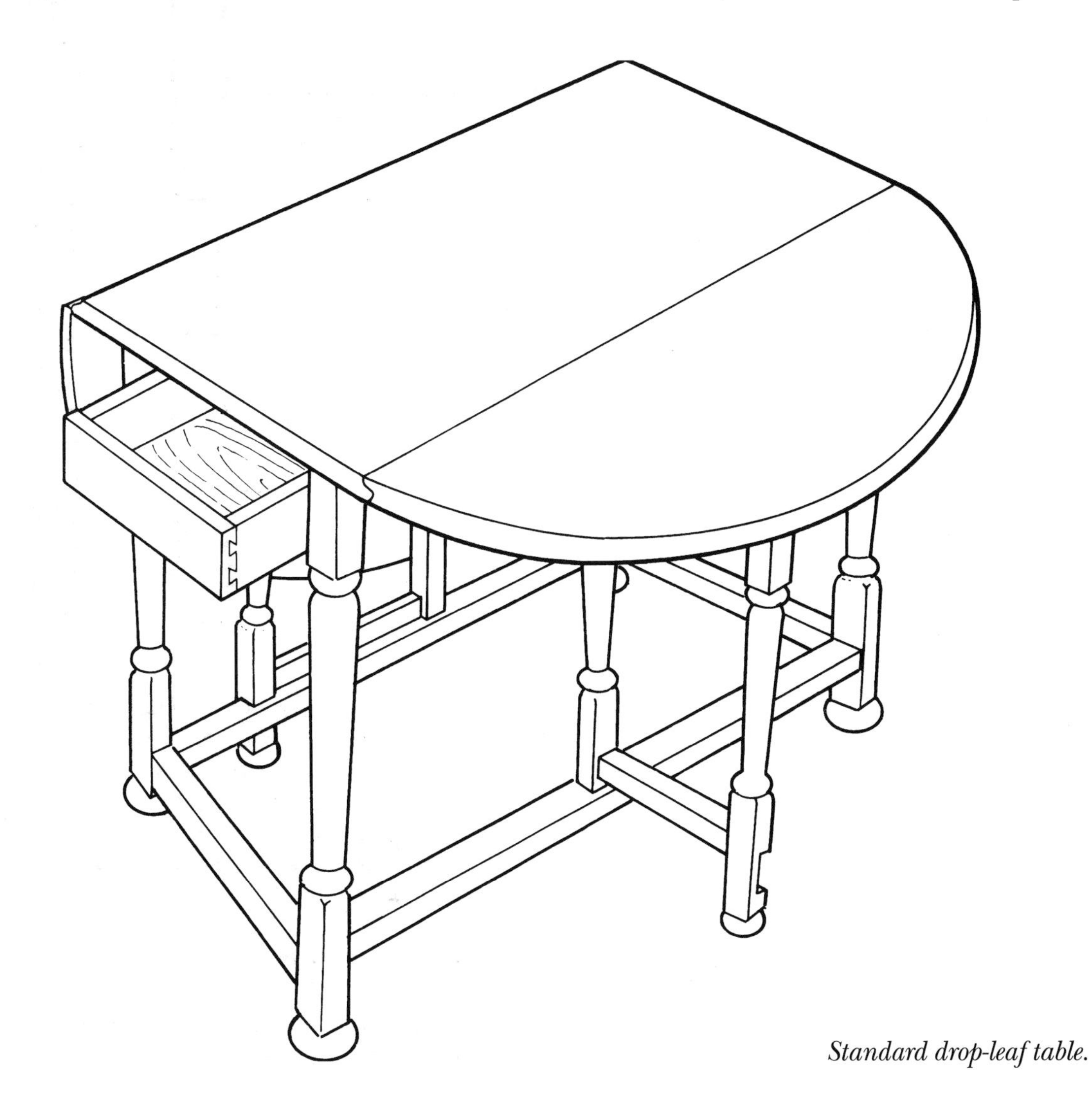

Standard drop-leaf table.

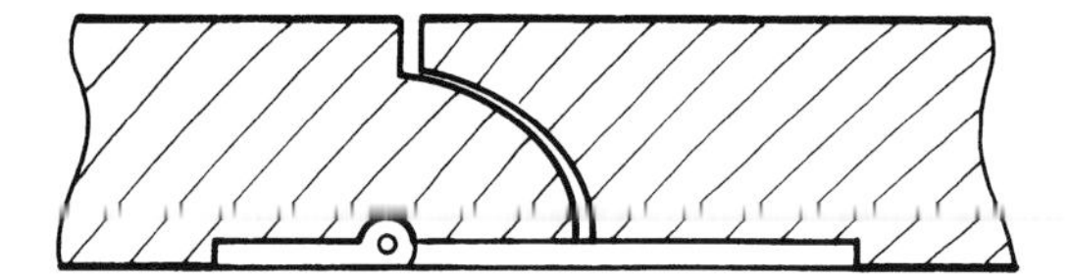

Drop-leaf table with rule joint.

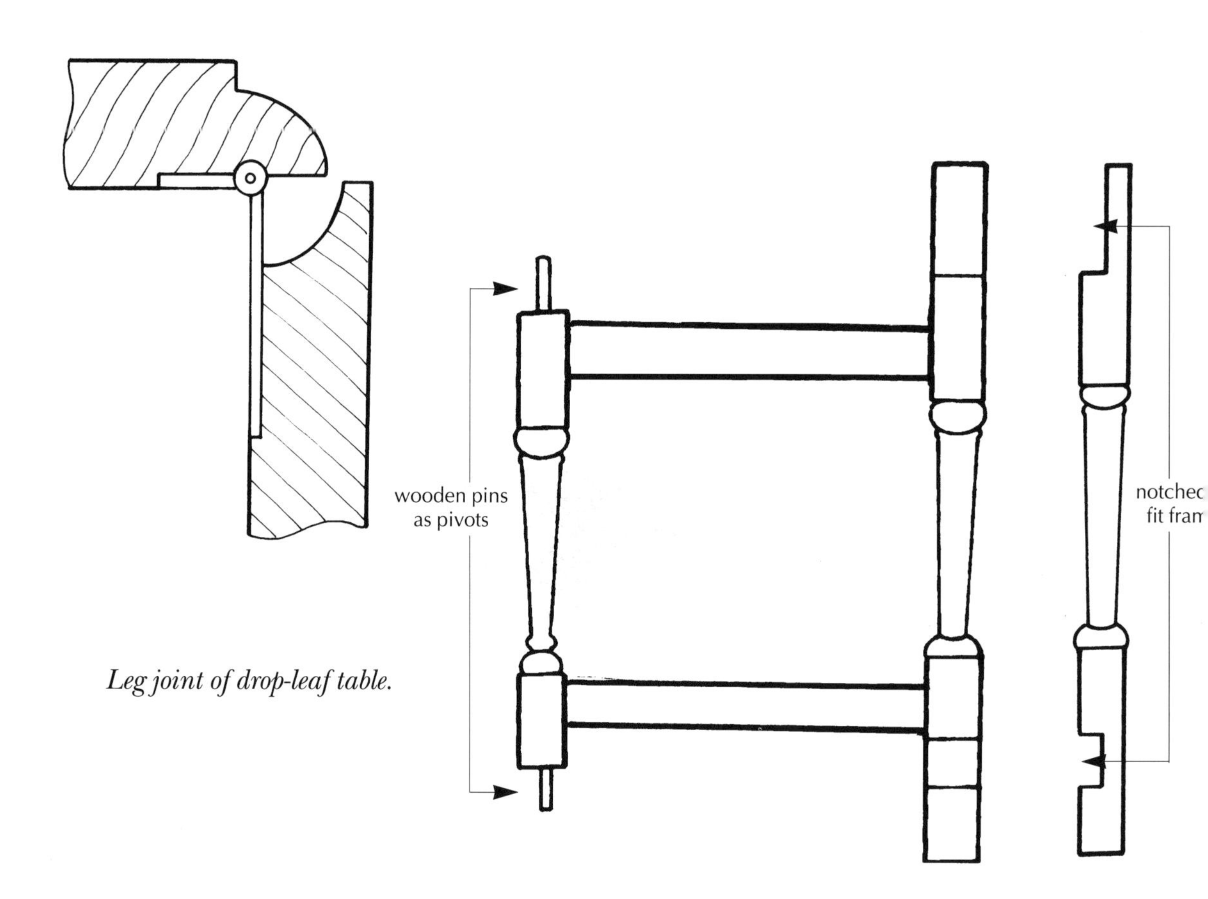

Leg joint of drop-leaf table.

with a detachable metal corner plate, which allows the frame to be held securely to the legs. The legs can be removed for transport or storage through the simple mechanism of a wing nut.

DROP-LEAF TABLE

It is possible to expand on the basic frame for a drop-leaf or gate-leg table. Here the main frame is constructed in the normal way but has an additional sub-frame consisting of two legs, one of which is pivoted with a dowel joint between the top rail and bottom stretcher. By lifting the table top to its horizontal position the sub-frame swings into a position that allows it to support the table leaf.

TRADITIONAL WRITING TABLE

With a traditional writing table the frame is again constructed in the standard way,

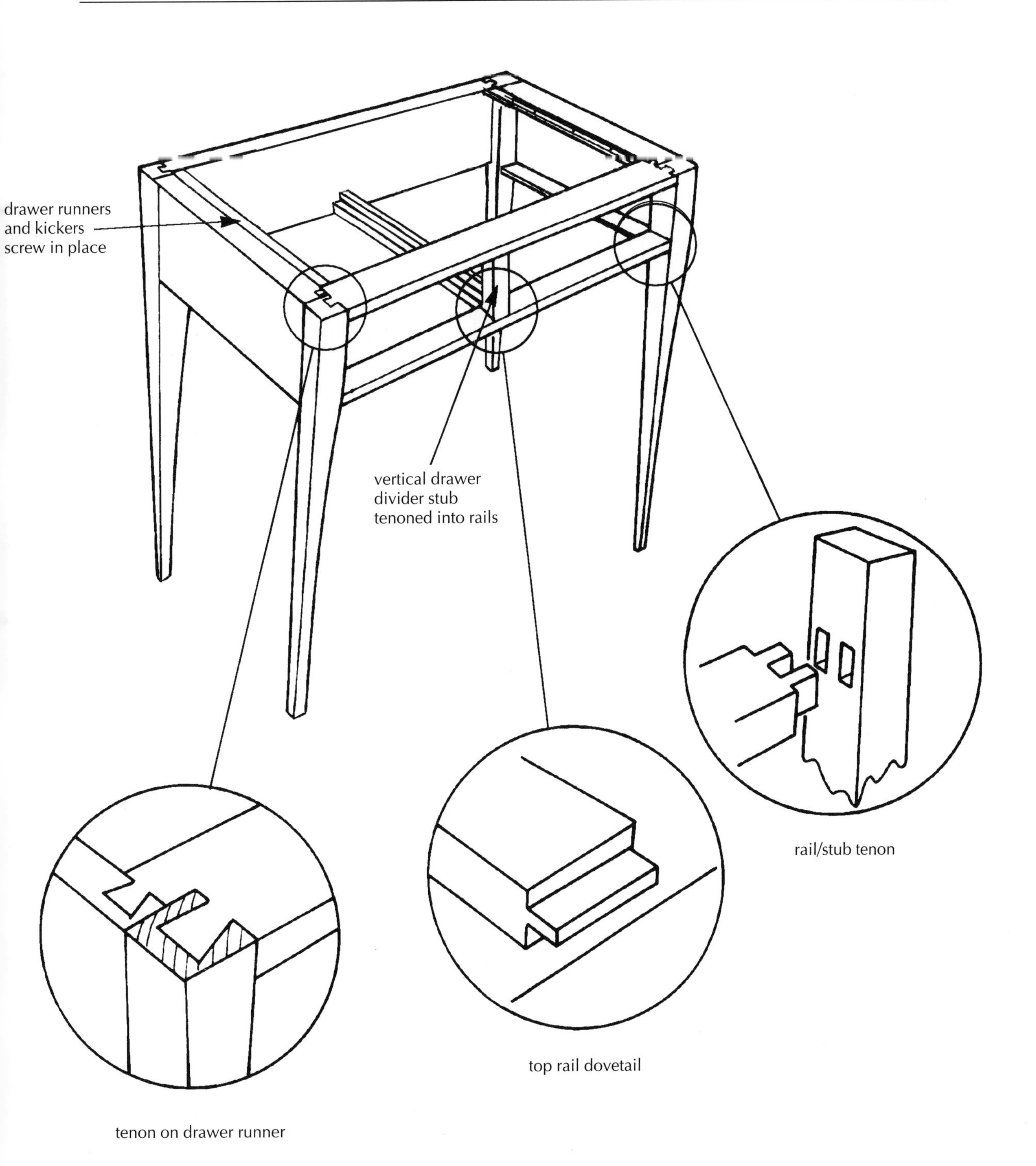

Writing table construction.

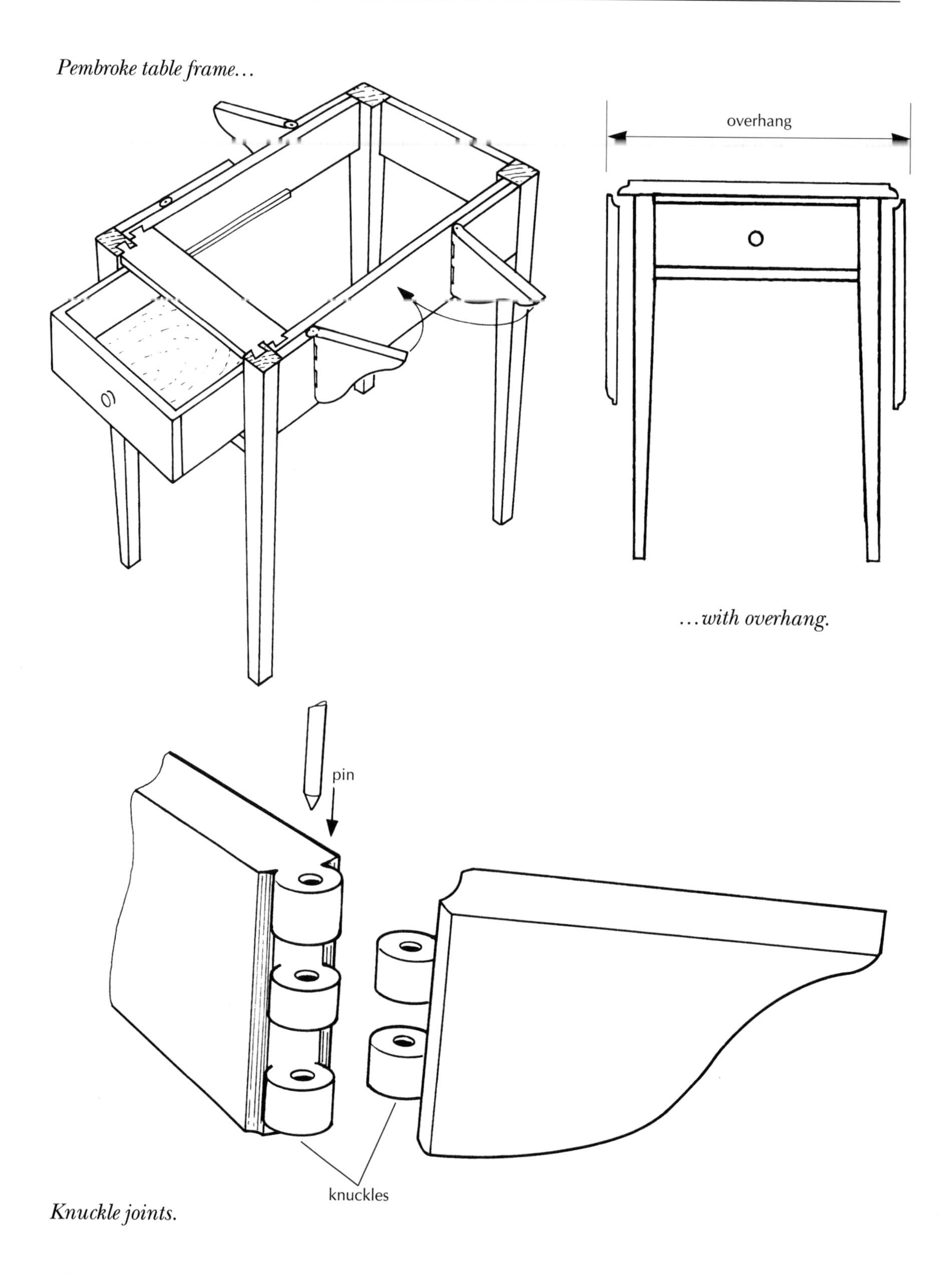

Pembroke table frame...

...with overhang.

Knuckle joints.

with the exception of the front rail. In this case there are two horizontal rails rather than one vertical rail. The bottom rail is held in position with a stubbed tenon and is sometimes called a bottom drawer rail. The top drawer rail is held in position by means of a lapped dovetail. The space between these rails is often occupied by one or two drawers.

PEMBROKE TABLE

A Pembroke table is a small rectangular table with drop leaves supported by brackets that pivot out from the front and back rails, rather than with a gate as in the case of the drop-leaf table. These brackets are commonly made of beech and have knuckle joints cut into their ends to facilitate the pivoting action. The table end houses a drawer that is fitted between the bottom and top drawer rails as in a traditional writing table.

CIRCULAR LEGS

Some table frames are made using circular legs to emphasize a particular design. The main frame adheres to the same principles that have already been outlined, but there are one or two modifications that need to be made in the jointing

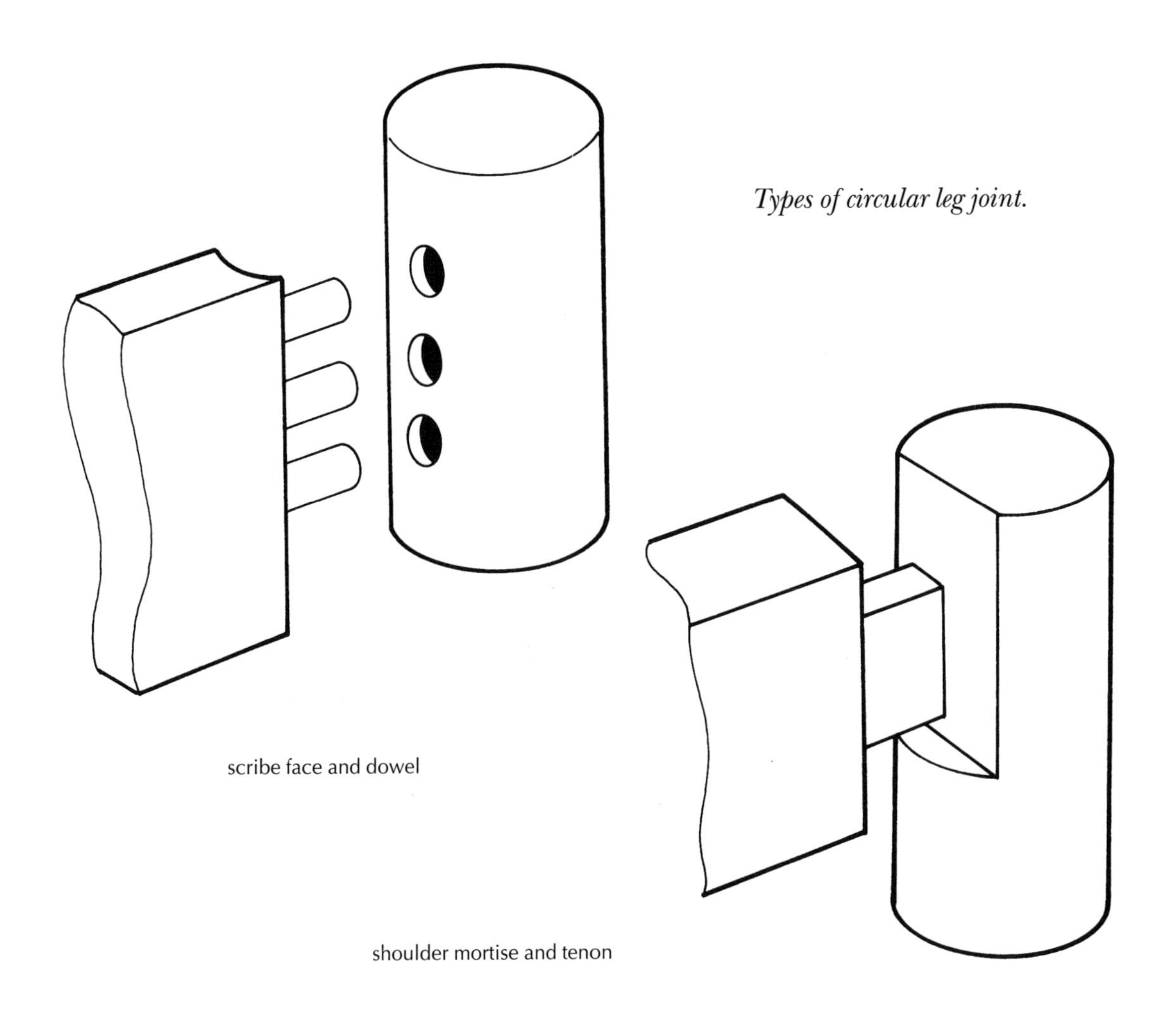

Types of circular leg joint.

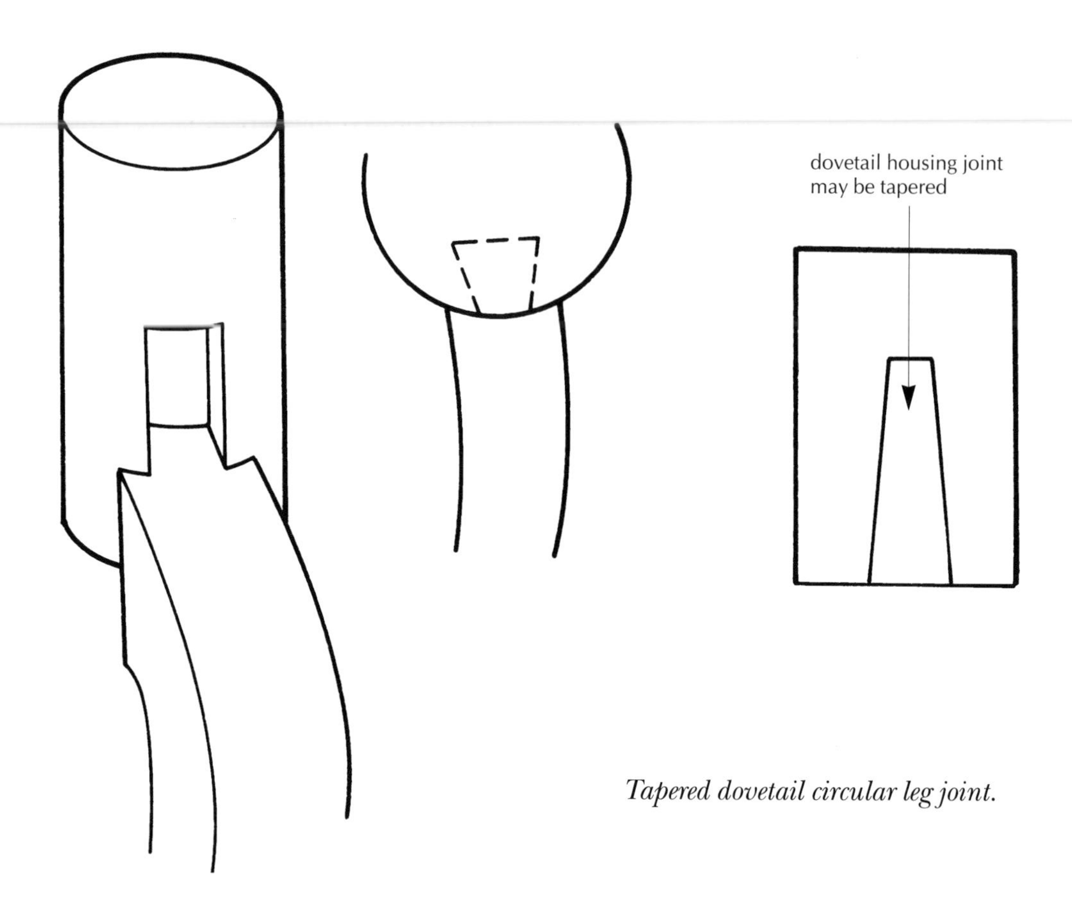

Tapered dovetail circular leg joint.

to accommodate this. One technique is to put square faces onto the circular legs where the rail is to be jointed, thus allowing the shoulders of the rails to be square. This allows for the fitting of a standard mortise and tenon or dowel joint. Another option is to scribe the shoulders of the rail to match the radius of the leg. This provides a higher quality finish, but involves more time and effort.

Tripod tables frequently have scribe joints. One of the best joints that can be employed to join the legs to the central column in this situation is a tapered dovetail housing joint. The tapering of the joint allows it to be easily tapped into place should it work loose and helps ensure a tight fit when cramping a rounded joint is difficult.

PROBLEMS

Any kind of construction operates to the same fundamental principles, whatever the piece of furniture. Once the required joints have been made it is simply a case of putting it all together in the same way as one would a three-dimensional jigsaw puzzle. Most problems that occur are usually the result of inappropriate construction that does not take account of the natural movement of solid timber or of inaccurate work. As it is a three-dimensional puzzle it is important that the piccc is joincd together in the correct sequence; perhaps the best way to illustrate this is to give some examples.

SIMPLE BOOK CASE

A standard book case is basically a box that houses a series of shelves in various formations and sizes. The side panels will be made of whatever materials are required and are of indeterminate size. The top panel will be joined to the side panels with a number of different joints:

- A butted dowel or biscuit joint. This is the quickest and cheapest joint to use and will predominate in most book cases.

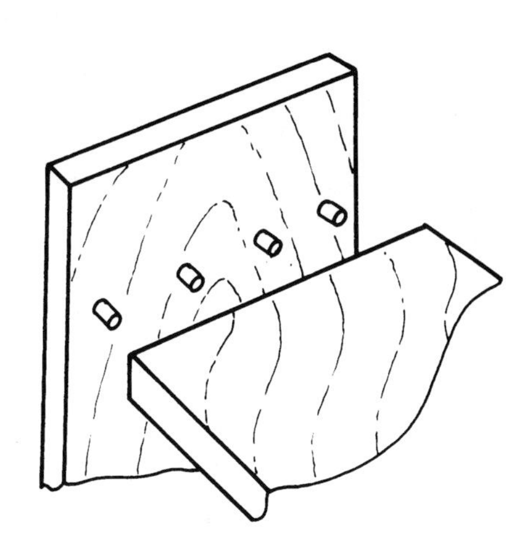

Dowel and biscuit joints.

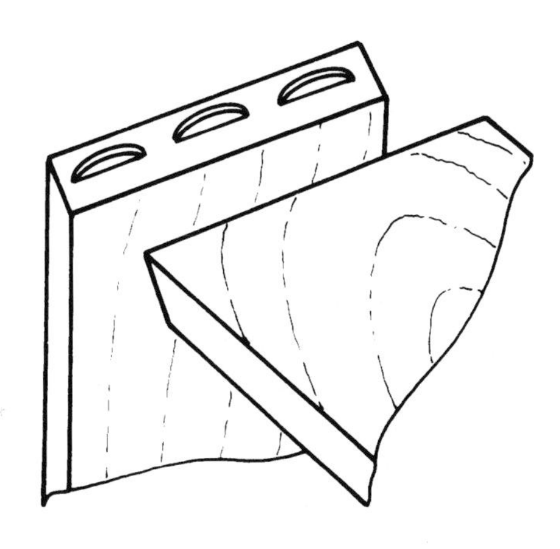

Mitred biscuit joint.

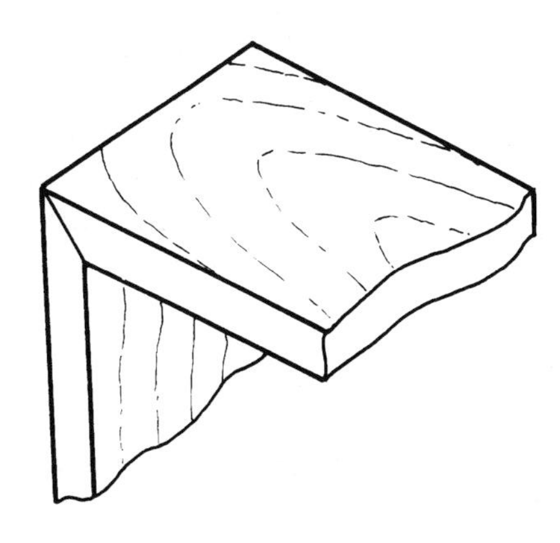

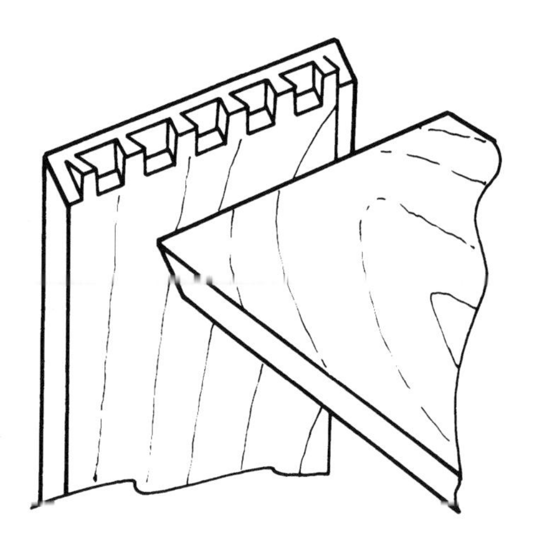

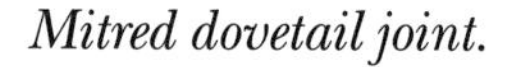

Mitred dovetail joint.

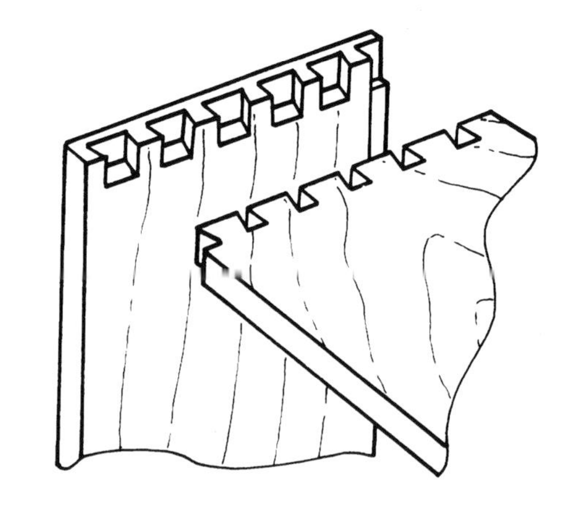

Lapped dovetail joint.

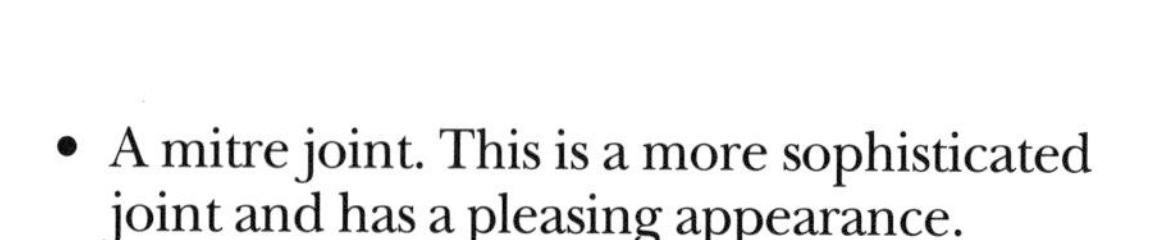

- A mitre joint. This is a more sophisticated joint and has a pleasing appearance.
- A secret mitred dovetail. This would be used on an expensive, more up market piece, largely because of the time element involved in the construction rather than the difficulty.
- A lapped dovetail. This is an option when a cornice or a false top is to be applied and is used for design rather than practical purposes.

The bottom rail can be fastened in the same manner as the top rail and a plinth applied afterwards. Alternatively the sides can continue past the bottom shelf, which can be fixed with a biscuit or dowel joint.

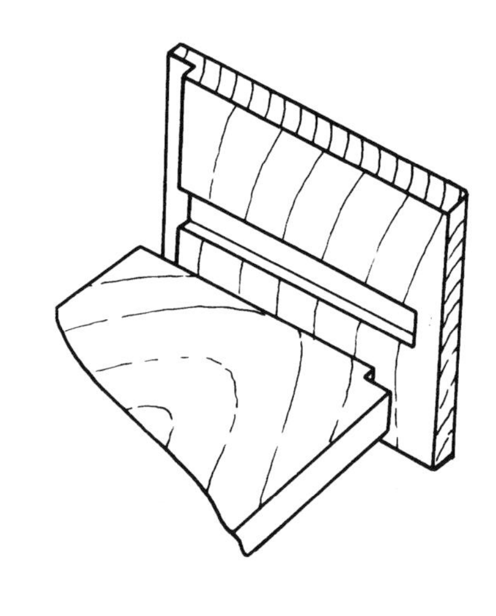

Stopped housing joint.

The shelves may be fixed or adjustable. If they are to be fixed it is useful to use a housing joint as in the bottom rail. If they are to be adjustable it is possible to purchase brass dowels. The back can be held in a groove with screws to lock it in place or if it is solid timber can be fixed in a rebate.

TRADITIONAL CUPBOARD

A traditional cupboard is constructed along the same principles as the bookcase. However, a bookcase tends to be narrower than a cupboard so there is less scope for movement than with the wider cupboard top. Also, the cupboard commonly has an overhanging top for ease of construction and as a design feature. Therefore place two horizontal rails on the front and back to secure the carcass in place and slot screw the top onto the panels. The jointing options are the same as with the bookcase and remain a matter of individual taste.

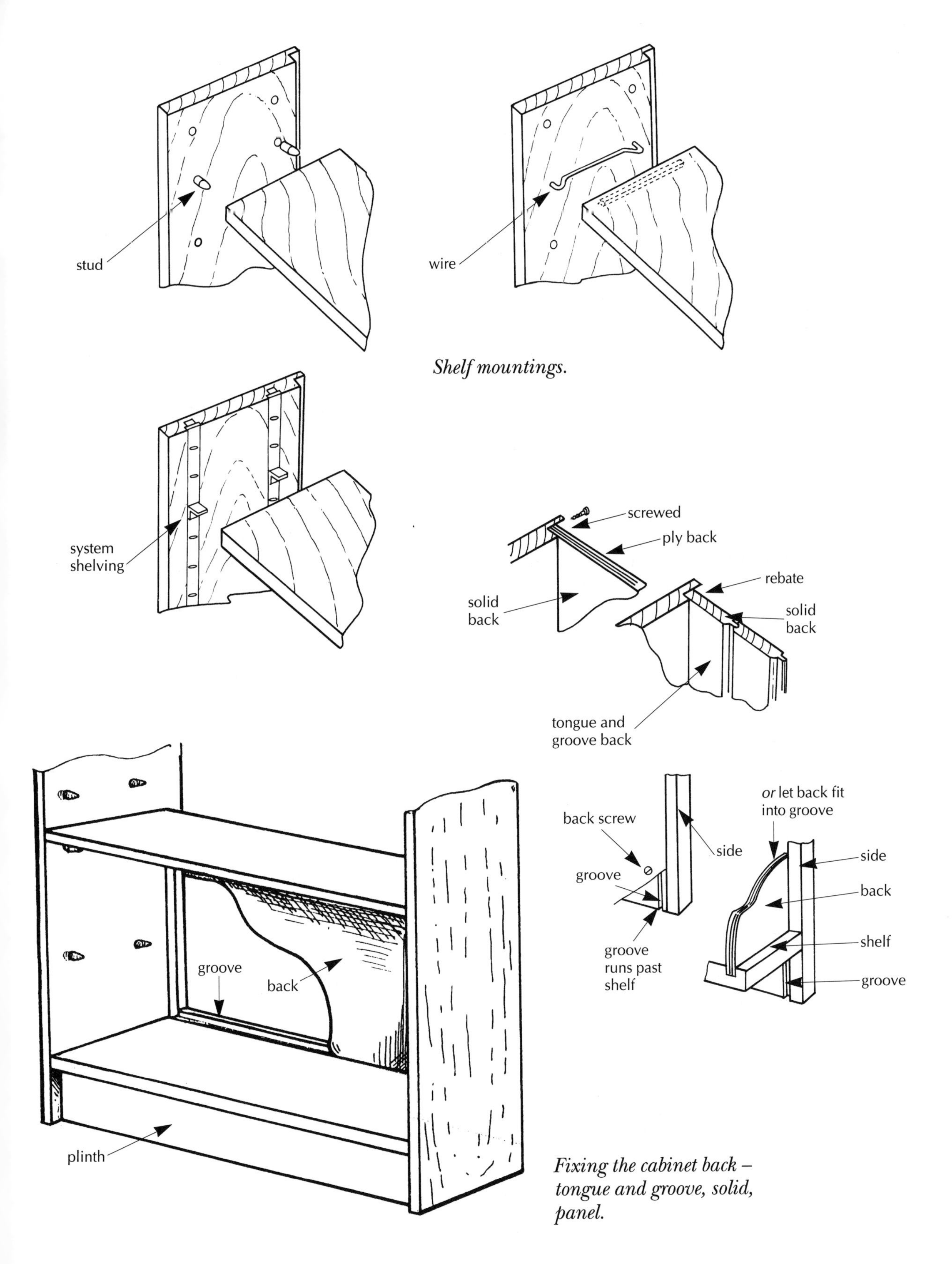

Shelf mountings.

Fixing the cabinet back – tongue and groove, solid, panel.

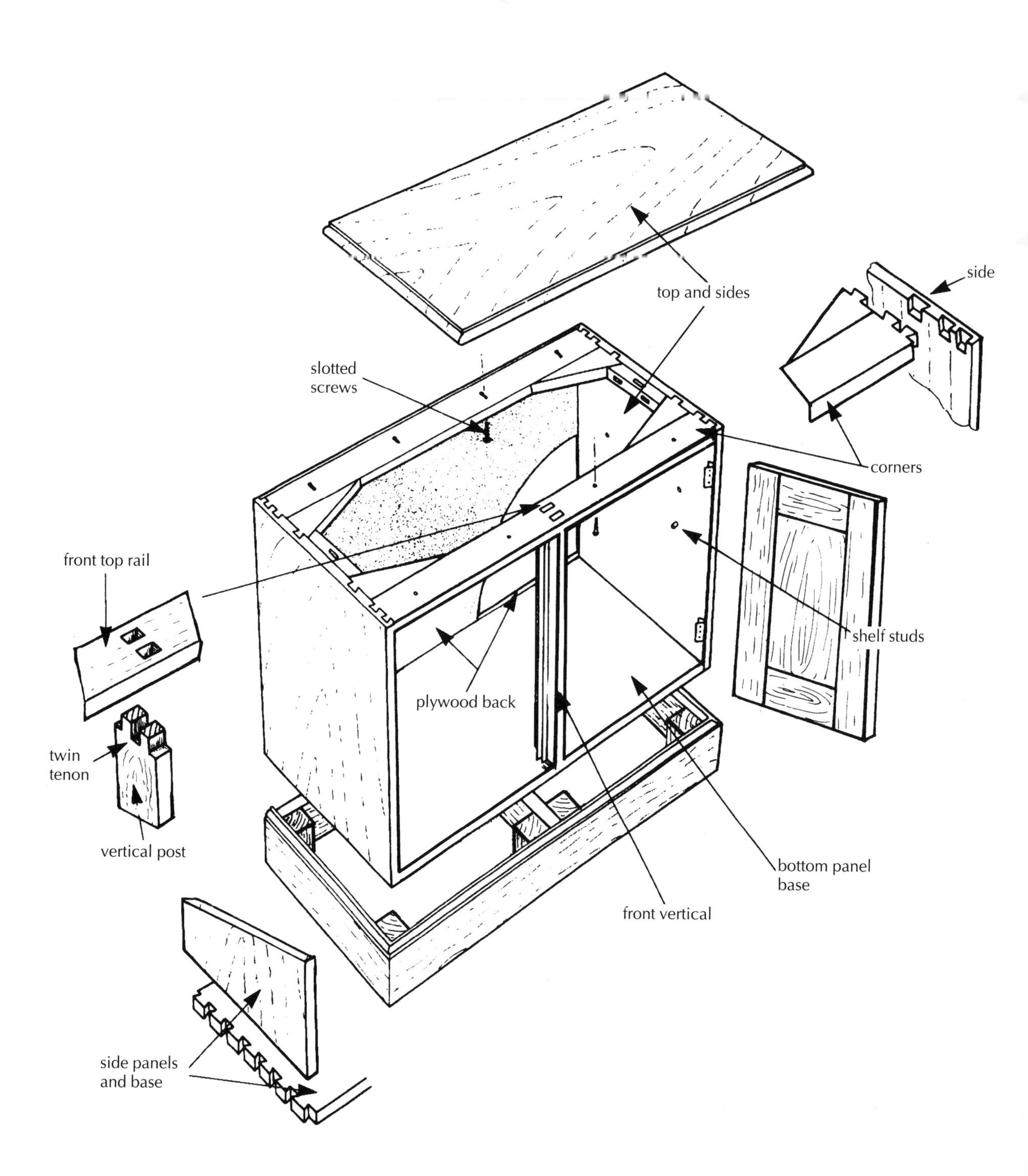

Traditional cupboard components.

CHAPTER **twelve**

Doors, Drawers and Locks

DOORS

Many types of door are used in cabinetry and the choice and style selected will depend on the design of the carcass and its function. The traditional door bears some similarities to a carcass back in that it is generally made up of a frame and panel construction and put together with mortise and tenon joints or dowel joints.

The construction of simple doors can vary. It is important that the rails and stile for all panels are square. A thinner door panel will be fitted into a

> **TIP**
>
> Remember to check the dimensions of the panels before placing them in the grooves and proceeding with the general assembly. They should always be worked with the same grooving cutter as the rails and stile.

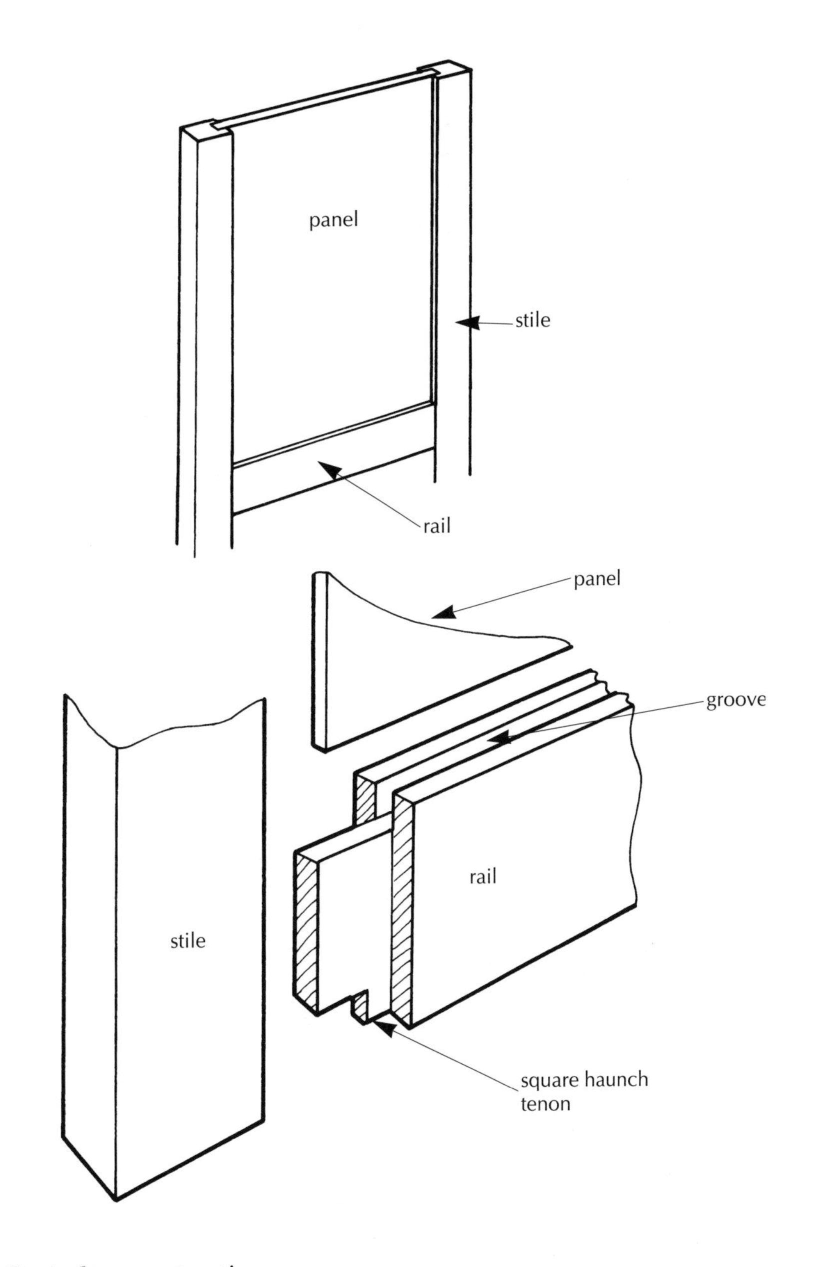

Basic door construction.

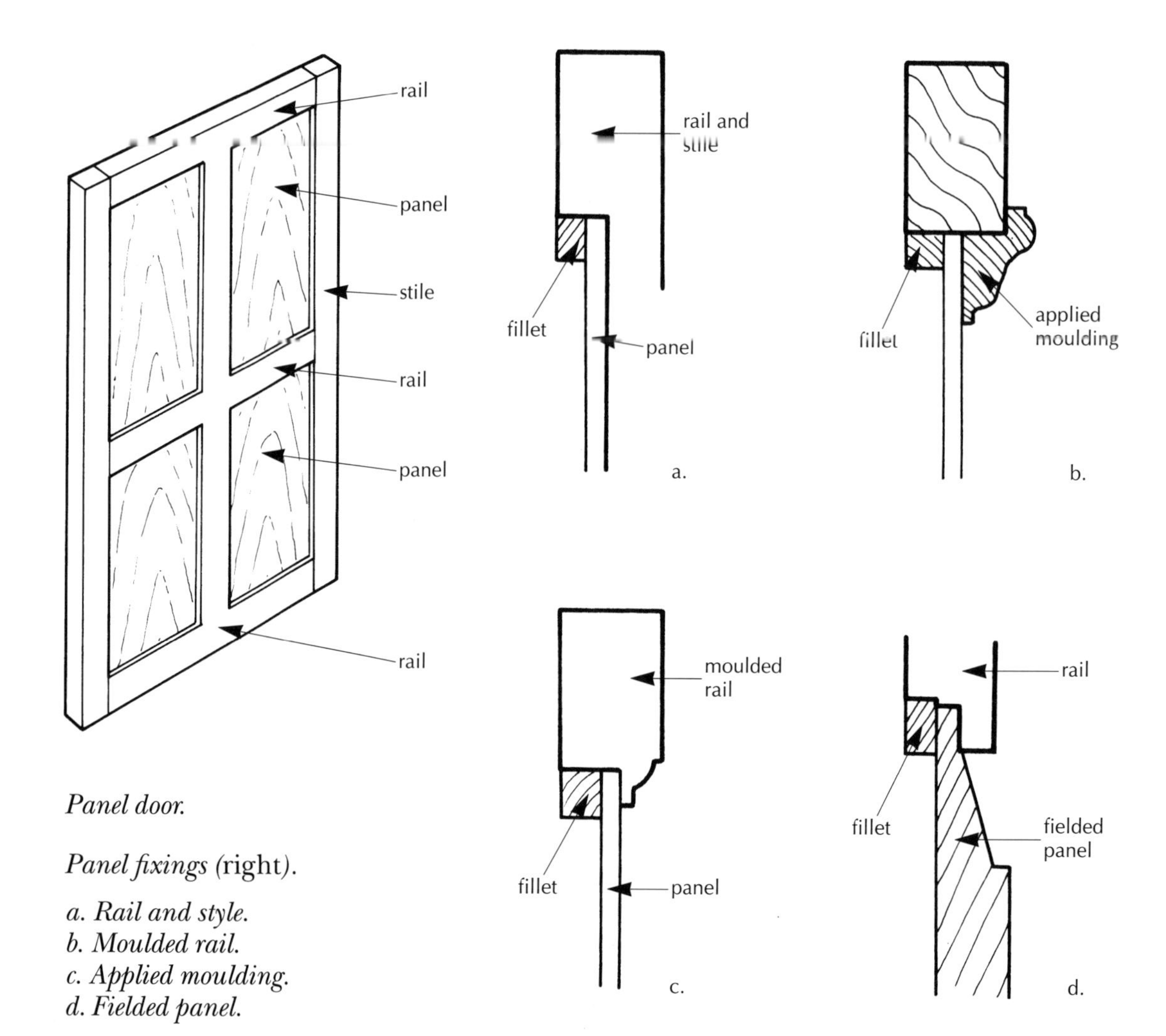

Panel door.

Panel fixings (right).

a. Rail and style.
b. Moulded rail.
c. Applied moulding.
d. Fielded panel.

rebate or groove. This will allow for movement of the solid panel as a consequence of shrinkage or expansion. It is essential that such a panel is loose within the groove and not glued into position. This is not applicable to panels formed of a synthetic material such as plywood as these are not subject to the movement found in solid woods.

DOOR FITTING SEQUENCE

It is important that doors are straight and square so that they fit comfortably into the carcass and operate smoothly. When fitting doors the following steps should be taken:

1. Using a tenon saw cut the horns on the stiles off and carefully plane flush with the rails.
2. Plane the hanging stile to fit the cabinet.
3. Plane to the correct width.
4. Plane the bottom edge.
5. Pack the bottom edge with a piece of veneer and plane the top edge.

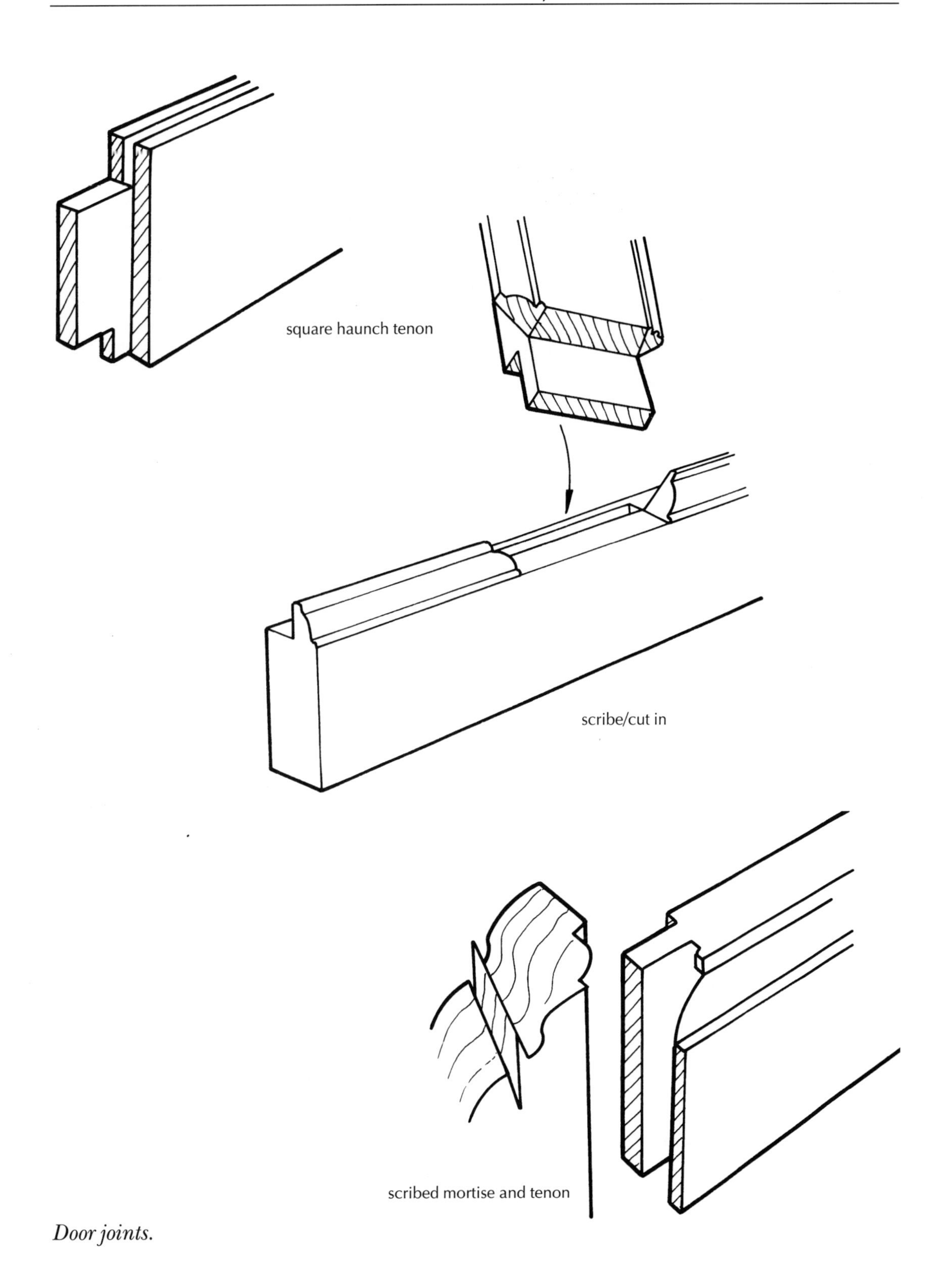
square haunch tenon
scribe/cut in
scribed mortise and tenon

Door joints.

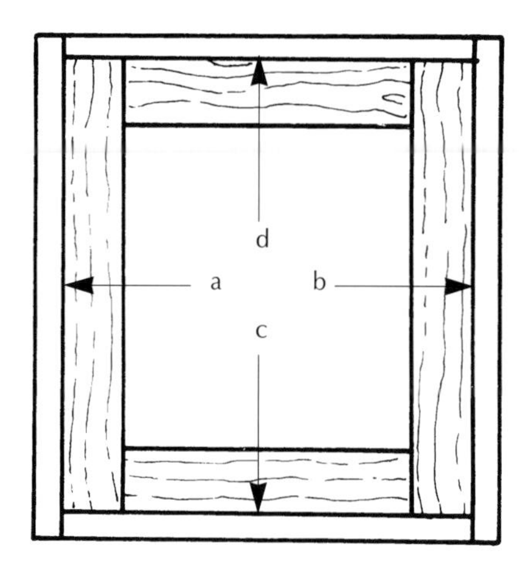

Sequence for fitting.

6. Check the clearance with a piece of veneer. This should amount to approximately 0.6mm (1/32in) all around.

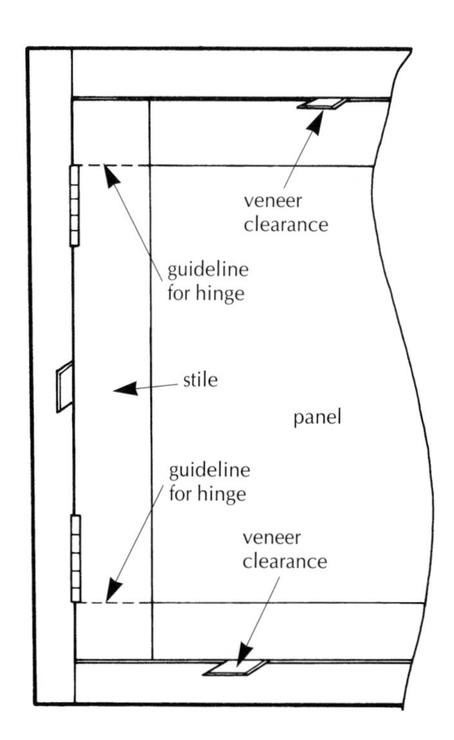

Checking clearances.

HINGES

Hinges serve two purposes. The first is practical and obvious – to allow for a door to be opened in the correct fashion. They therefore need to be accurately fitted and well prepared. However, hinges can also serve a decorative purpose and can augment the aesthetic quality of a piece most effectively. The type of hinge fitted and the way in which it is employed will depend on the requirements of the item being made, but it is an important feature that should not be overlooked.

The hinge can be made from any number of materials, but one of the most common and useful materials in both design and practicality is brass. There are a great many options to choose from, some being relatively specialized and expensive, but below are descriptions of some of the more common types of hinge.

The *concealed cabinet hinge* is often used for the doors of kitchen cupboards aligned one against the other along a wall. It prevents one door slamming into its neighbour. A hole is generally drilled into the door, into which is fitted a circular boss. The base plate screws into the cabinet. The *cylinder hinge* is commonly used in concertina doors because it is designed to allow doors to be opened to 180°. Holes are drilled into the wood, so once fitted these hinges cannot be seen when the doors are closed.

The *centre hinge* is a simple hinge that is recessed into the edge of the door and is very difficult to see when the door is closed. The *backflap hinge* is broad and usually made of solid brass. Its most common purpose is to attach bureau flaps. The *table hinge* is very similar in design and appearance to the backflap hinge. Again made of solid brass it is particularly useful for fixing fold down table flaps. The longer leaf of the hinge is fixed to the flap.

The *butt hinge* is the staple hinge of the cabinet maker. It is nearly always made of

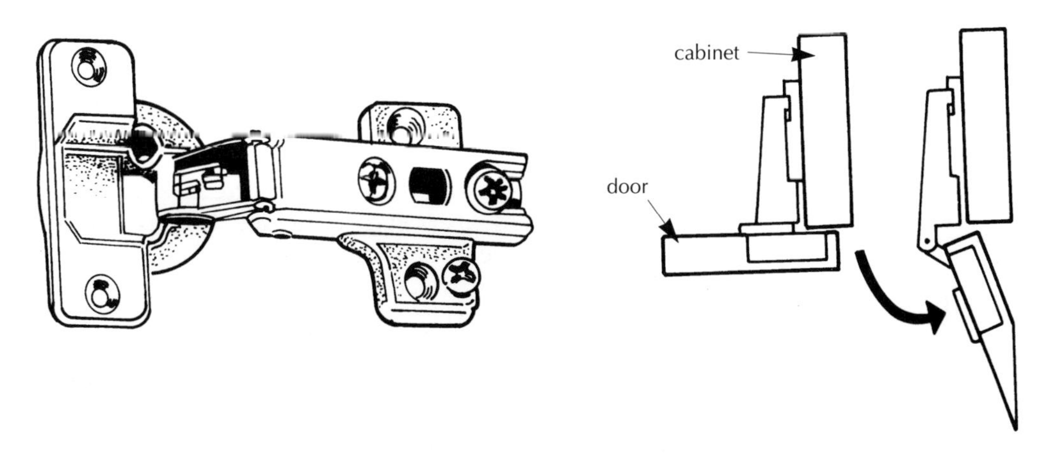

Concealed cabinet hinge.

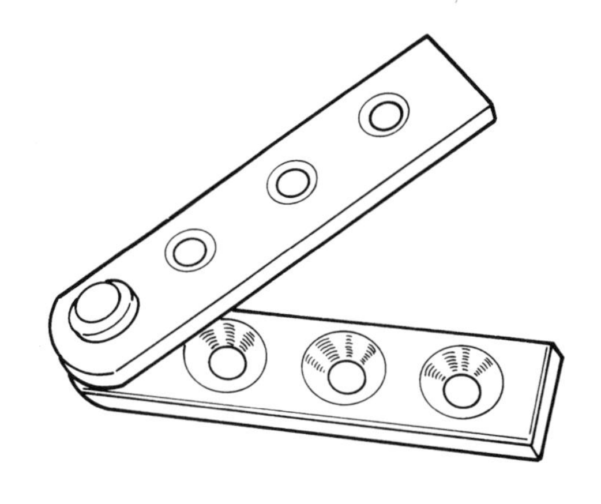

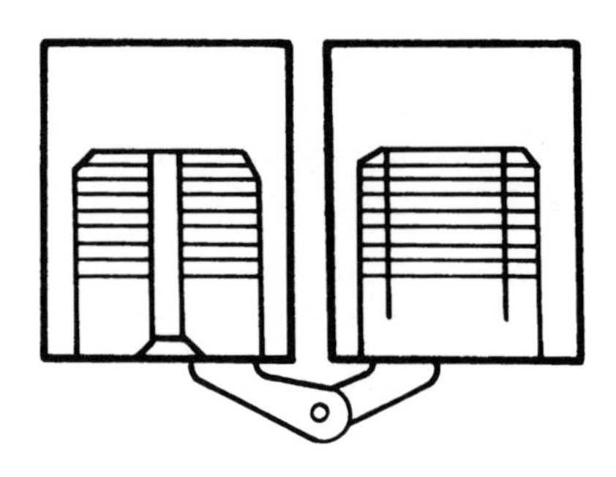

Cylinder hinge.

Centre hinge.

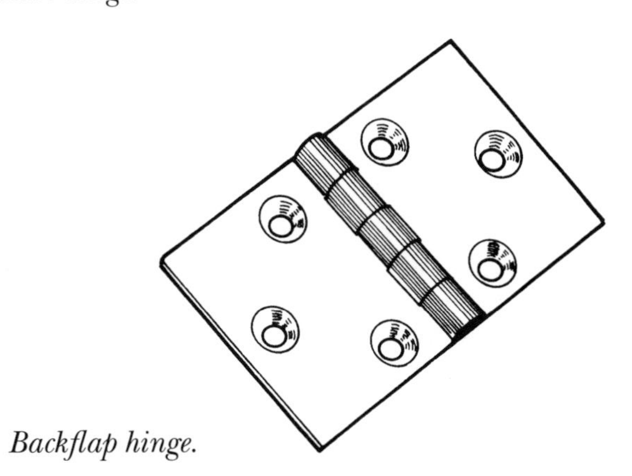

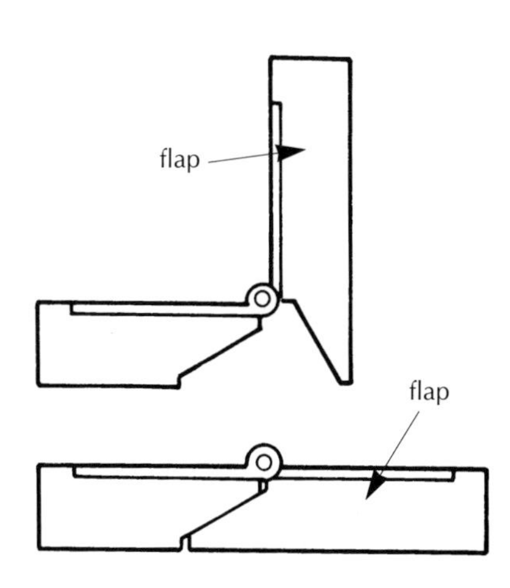

Backflap hinge.

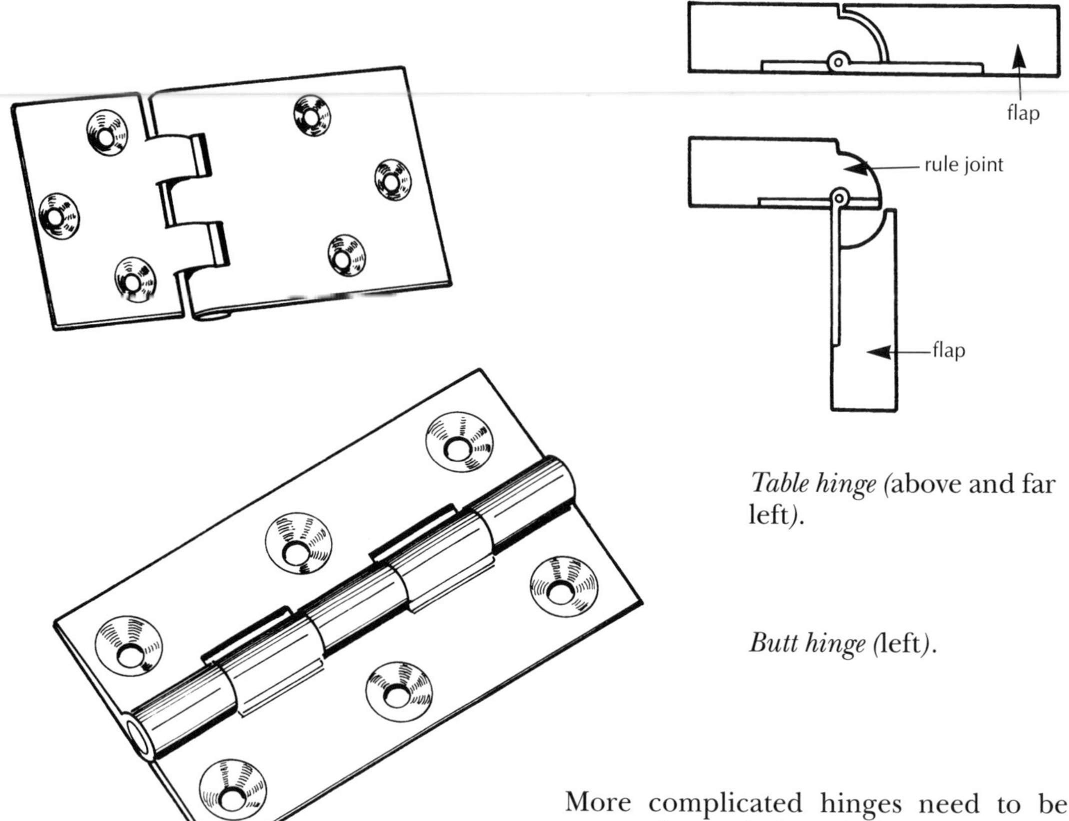

Table hinge (above and far left).

Butt hinge (left).

brass and comes in a small range of sizes with the widest being particularly suitable for larger pieces such as wardrobes. The narrower versions can be used in smaller pieces.

Each hinge that is fitted to any type of furniture has to be individually assessed. In the case of simple butt hinges there is no need to be perfectly precise, so long as there is sufficient support for the door or the flap. However, it is important to take into account the appearance of the hinges – if haphazardly placed they can make the whole piece seem to be out of perspective.

The hinges should be recessed equally. Allow for a smooth fit so that there is no visible gap where the knuckle throws the hinge or the door binds into the carcass.

More complicated hinges need to be measured precisely. If the hinges are hand-made they will require individual marking out, with each hinge on each piece having to be considered separately.

Mark the starting position of the hinge with a knife or scalpel (1). It is customary to have this level with the middle edge of the rails. Hold the hinge firmly in position and mark out the length of the hinge (2). The marking gauges should be set so that the flange is moved to the centre of pin (a), then to the centre of the knuckle (b). Where the hinge is to be located entirely in the door, it should be set to the total knuckle thickness (c). Mark out using the gauges the square and a marking knife. When it is accurately marked out cut precisely to the mark with a dovetail saw or chop the fibre with a chisel. Clean out the waste with a chisel and make note of the taper to the back edge (d).

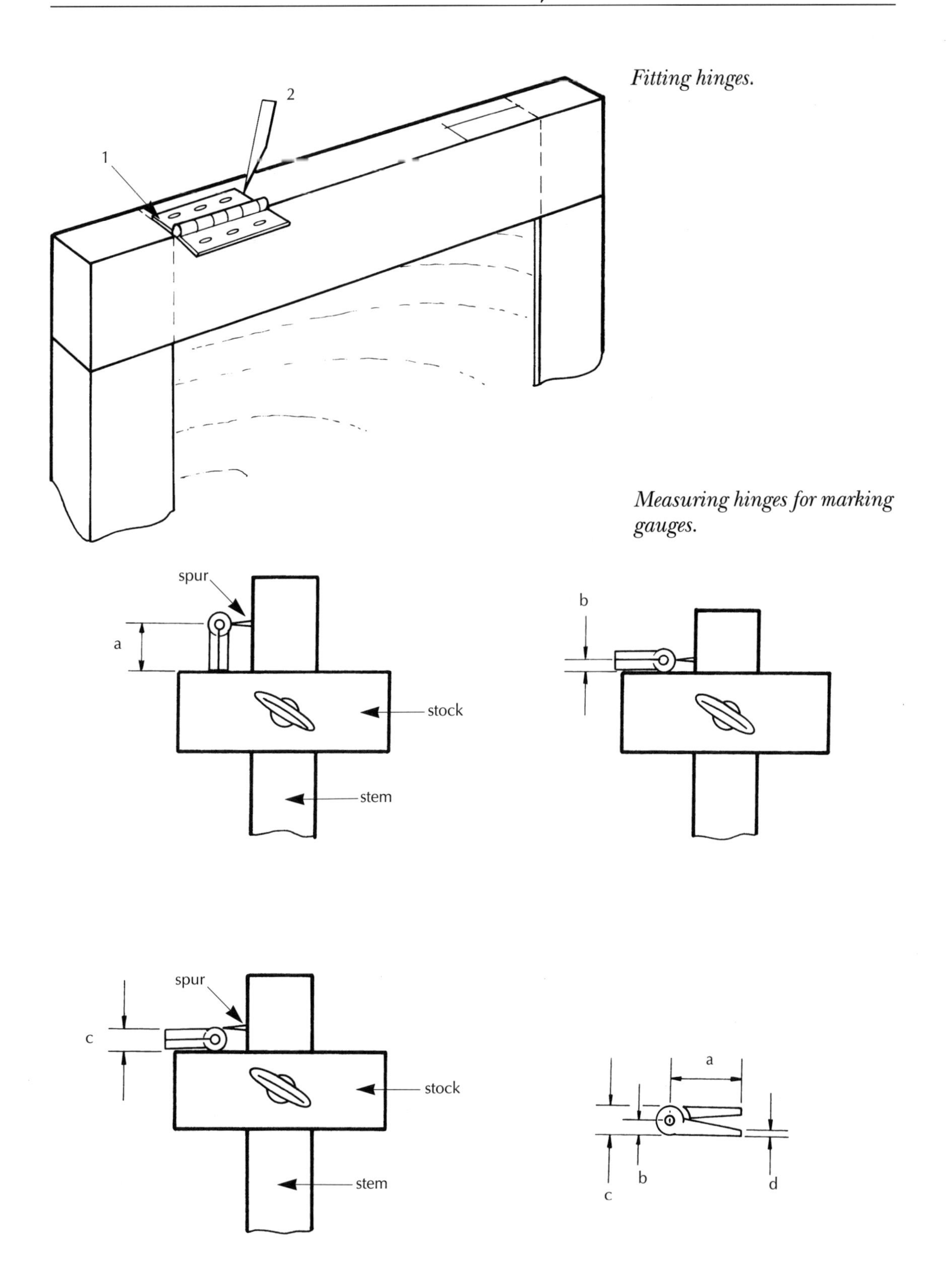

Fitting hinges.

Measuring hinges for marking gauges.

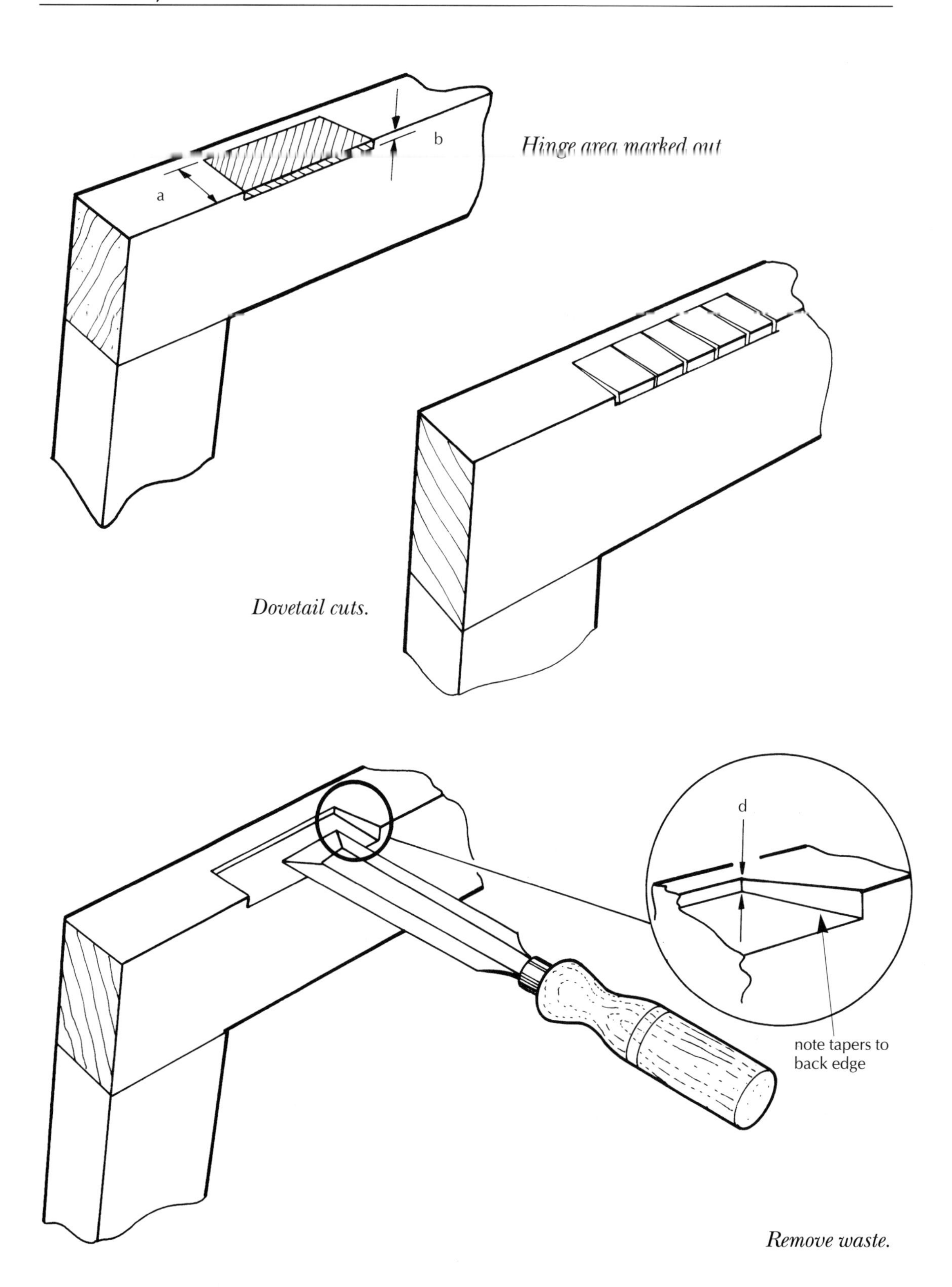

Hinge area marked out

Dovetail cuts.

Remove waste.

TIP

Use steel screws for the major part of the hinge fitting process because brass screws occasionally have their heads broken off if used in a wood that is particularly hard. Removing the thread that has broken from the head is an irritating waste of time. When it is time for the final assembly the steel screws should be replaced with brass screws. Remember to make sure that the countersunk hole is deep enough to take the screw head.

Now check that the hinge fits. The vast majority of traditional doors and boxes are fitted with butt hinges. Only the best solid drawn brass butts should be used as cheap hinges often fatigue and break over time. When you are certain that the hinge fits correctly, screw it into place with one screw initially and test the door. It may be that the door needs some slight adjustment with a plane to facilitate a perfect fit. Modify as necessary and complete the job by fitting in the remaining screws.

On high-quality work it is common practice to enhance the finished effect of the hinges by cleaning them up. This is easily done by working through the grades of emery cloth and by buffing them up to a mirror finish with metal polish before the final fit.

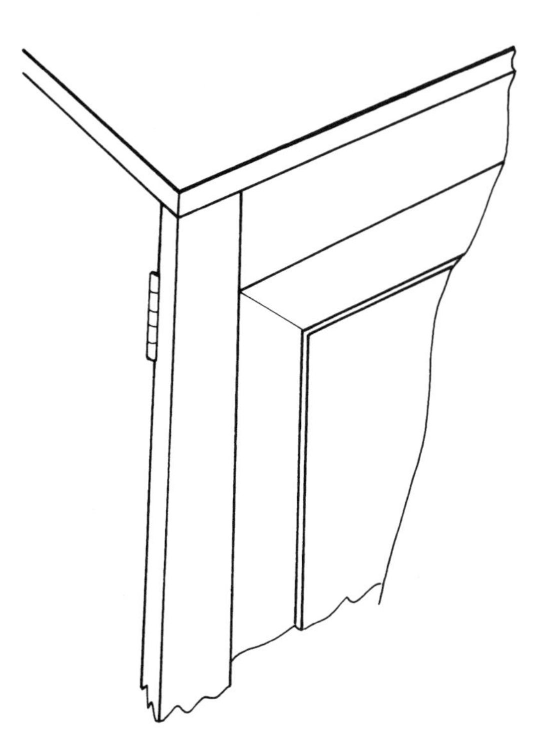

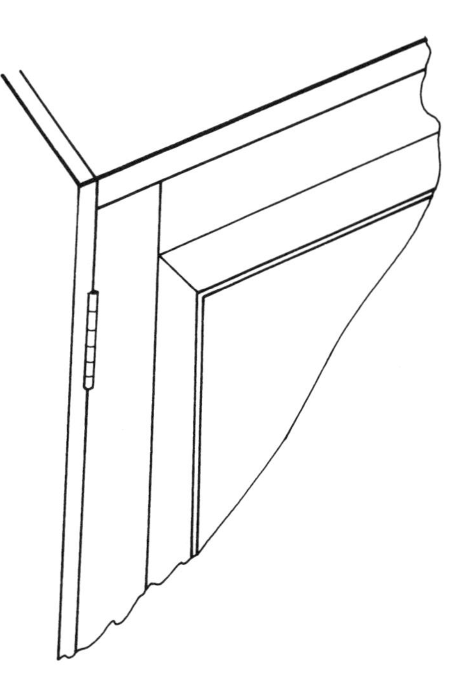

Door hinge positions.

STAYS

Stays are used to limit the swing of a door and are also used to support a fall down flap or prop up box lid. The stays are normally made in pairs and are designed with plates at each end for fixing. They are usually set at right angles to each other. It is important that the stays are of sufficient strength to carry out their intended task; a broken stay can lead to a damaged door at a later date.

DRAWER WORK

Drawers are fitted to desks, cabinets, chests of drawers, tables, wardrobes and many other items. As with framework, there are standard principles that should not be deviated from to any great degree, although there will be a need for minor adaptations.

The preparation of the drawer housing should be extremely thorough otherwise the drawer will not run smoothly or open correctly. The drawer components should fit perfectly and be tested prior to the joint being cut.

DRAWER RUNNERS

The selection of the material for the runners should be carefully considered. There is a fair amount of choice, but it is often wise to select a timber that is of a softer hew than the material making up the drawer itself – it is easier to replace worn runners than to replace a drawer that has been worn down by runners constructed of a harder timber.

> **TIP**
>
> Teak is a very suitable material for drawer runners, as it is naturally greasy and will lubricate the running action of the drawers over the rails.

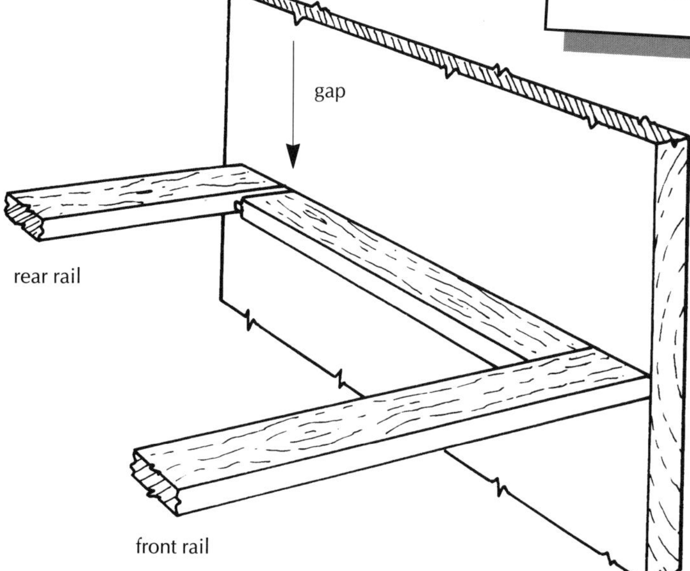

Drawer runner.

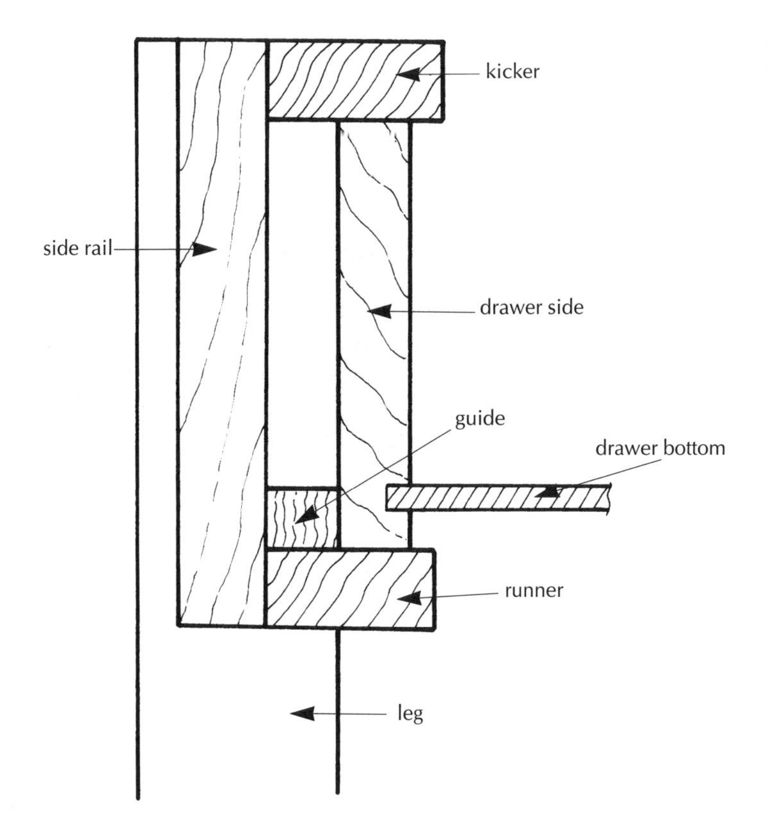

Side view of drawer and runner.

On a solid wood carcass there is no drawer guide, only a runner as the carcass side acts as the guide for the drawer. On a framed cabinet or table it is necessary to fix a guide either to the rail or the drawer runner because there is no side to guide the drawer.

CONSTRUCTING THE DRAWER

First plane all the timber to fit the openings. Start with the drawer front so that it fits and can be manipulated as a smooth, tight fit with no accompanying rattle. Now plane the drawer sides so that they run smoothly between the runner and the kicker. There should be no vertical slop. Cut the back of the drawer to the correct length so that it slides in. The timber is now prepared and it is possible to start marking out the joint. The front joints are always lapped dovetails and the back joints are through dovetails. They are marked out in the same way as described in the chapter on dovetails.

The next job is to fix the drawer bottom, and there are a number of ways of doing this. The best quality drawers normally have drawer slips. The upper surface of the drawer bottom runs level with the top of the drawer slip. The drawer bottom is made of Lebanon cedar with the length of the grain running from side to side. The bottom is attached to the back drawer rail with screws fitted through slots cut into the drawer bottom to allow for movement. On cheaper drawers a plywood bottom is simply slipped into grooves in the sides and drawer front and secured to the back drawer rail with screws.

Traditional drawer construction.

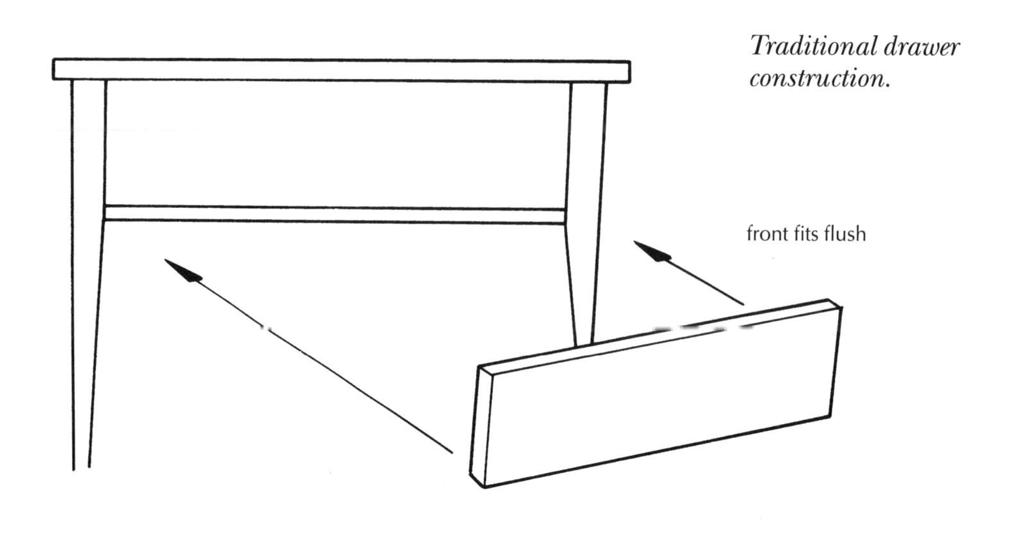

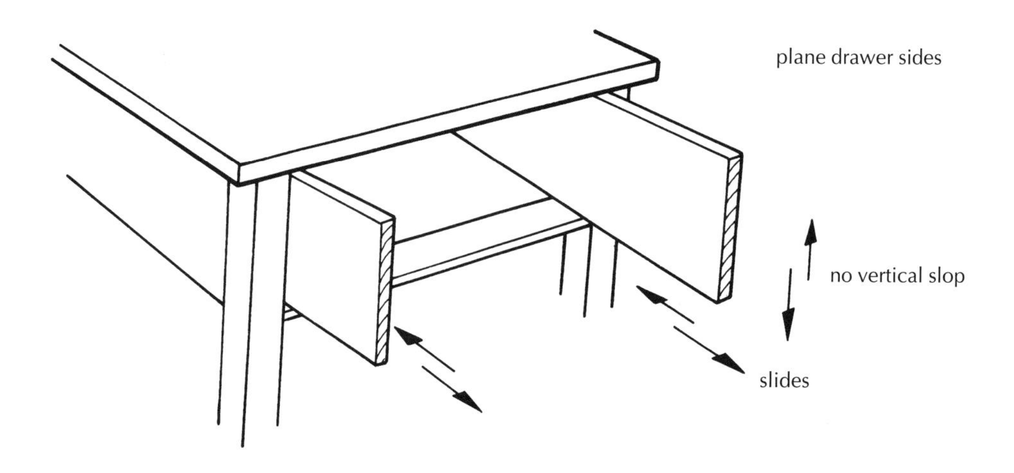

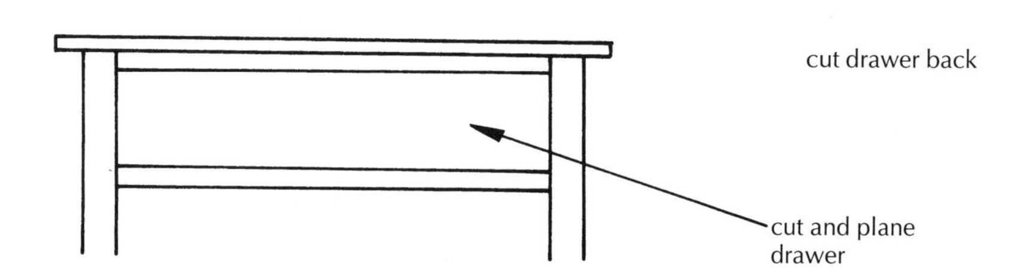

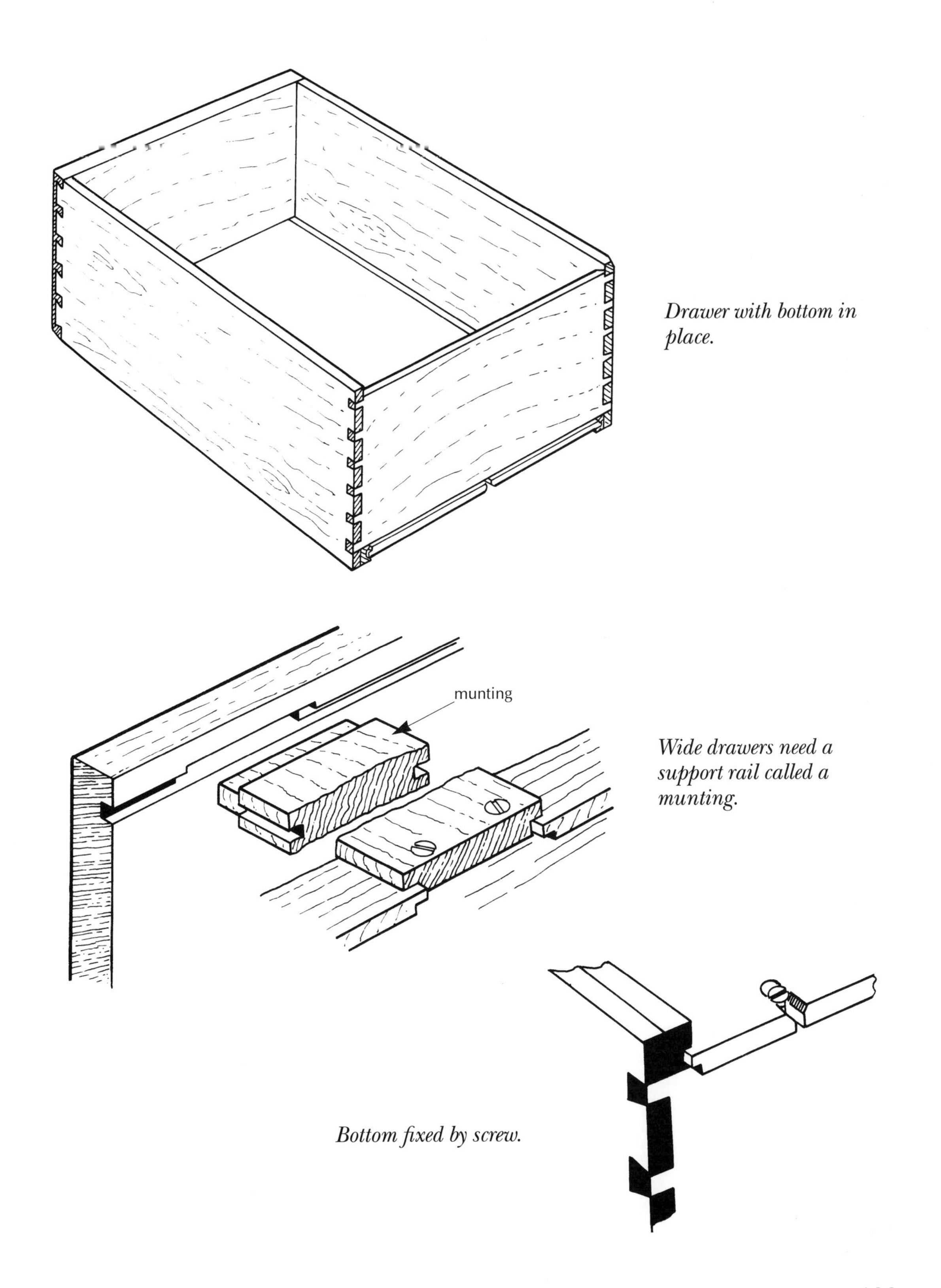

Drawer with bottom in place.

Wide drawers need a support rail called a munting.

Bottom fixed by screw.

LOCKS

There are many different fixtures and fittings that could be appended to furniture, but the lock is by far the most common. The lock has a practical purpose, but can also be an appealing decorative device, adding charm and character to a piece of furniture.

There was a period in the history of furniture making when almost every piece of furniture had a lock of one type or another. There were locks designed to keep things in and more importantly locks with the sole purpose of keeping people out. Part of the reason for this was the relative frailties of some of the catching systems employed, but by far the most important factor was mistrust of domestic servants. The most enthusiastic of the lock installers were by far the Victorians who took great exception to their staff sampling the vintage brandy or filching the odd coin. Locks were even fitted to drawers that were designed to store clothing – presumably the prudish Victorians wanted to keep their intimate garments from prying eyes. But whatever the motivation behind all this security, the demise of the domestic servant and the rise of mass produced furniture has put a stop to this trend, which is rather a shame.

There are occasions when a lock rightly forms an integral part of a piece of furniture. For these pieces it is invariably the decorative qualities of the lock that appeal, for it is no longer a protection against the modern burglar. Indeed it is probably wise to keep the key in the lock to prevent an avaricious thief from extracting revenge on the furniture or being drawn to it in the belief that valuable items must be stored inside.

TYPES OF LOCK

While there are some drawbacks in fitting locks to furniture – the main one being the time it takes – there is no doubt that a good, well fitted and sited lock enhances a quality piece. The *straight cupboard lock* is a double-handed lock. It is screwed directly into the inside face of the door without the necessity for any cutting away. It has the advantage of being easy to fit, but it is not the most attractive of locks and can seem obtrusive.

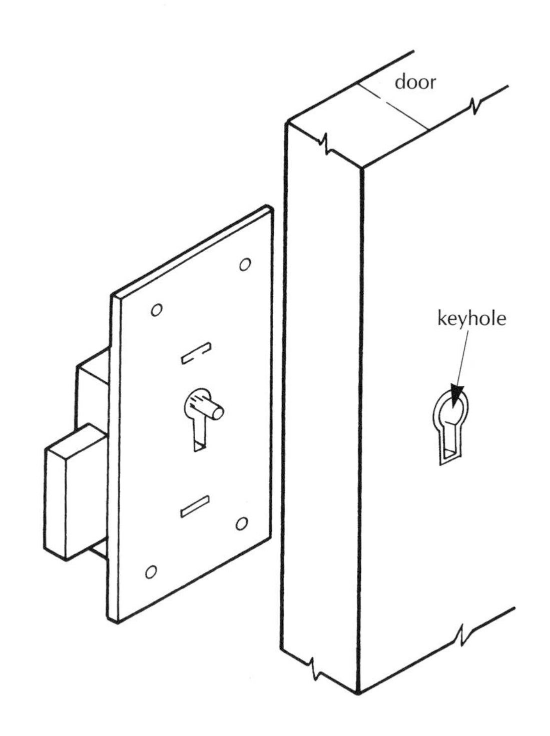

Straight cupboard lock.

The plate and body of the *cut cupboard lock* are set into the back of the door and must be bought as a left- or right-handed lock according to requirements. The reason for this is that the bolt can shoot only one way. The *mortise cupboard lock* also has to be bought in left- or right-handed form. It presents a neat and unobtrusive finish, but it is important that the doors are hung absolutely true and square if it is to be effective.

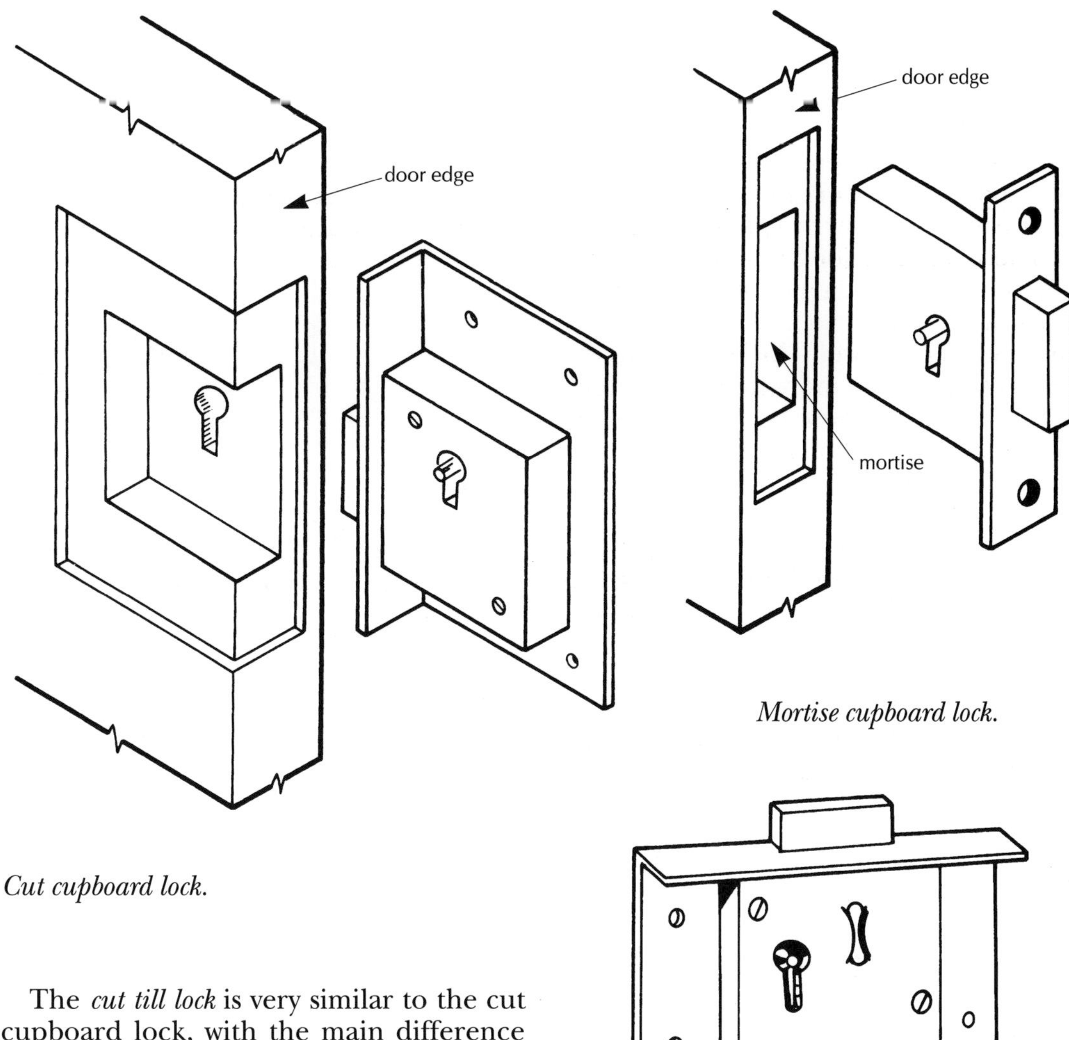

Mortise cupboard lock.

Cut cupboard lock.

Cut till lock.

The *cut till lock* is very similar to the cut cupboard lock, with the main difference being the configuration of the keyhole. It is possible to obtain these locks with two- or three-way keyholes to give a good measure of variety. The body and plate of the *cut wardrobe lock* are screwed into the back of the door in the same way as the cut cupboard lock and should be similarly fitted and handled.

The *box lock* is very different to the locks that have previously been described. It requires its own striking plate because of the action of the plate, which shoots upwards and then moves sideways to lock the door by hooking onto the plate.

FITTING LOCKS

Fitting locks is an intricate and time-consuming operation and needs attention to detail. It is partly the reason why locks on furniture have gone a little out of fashion.

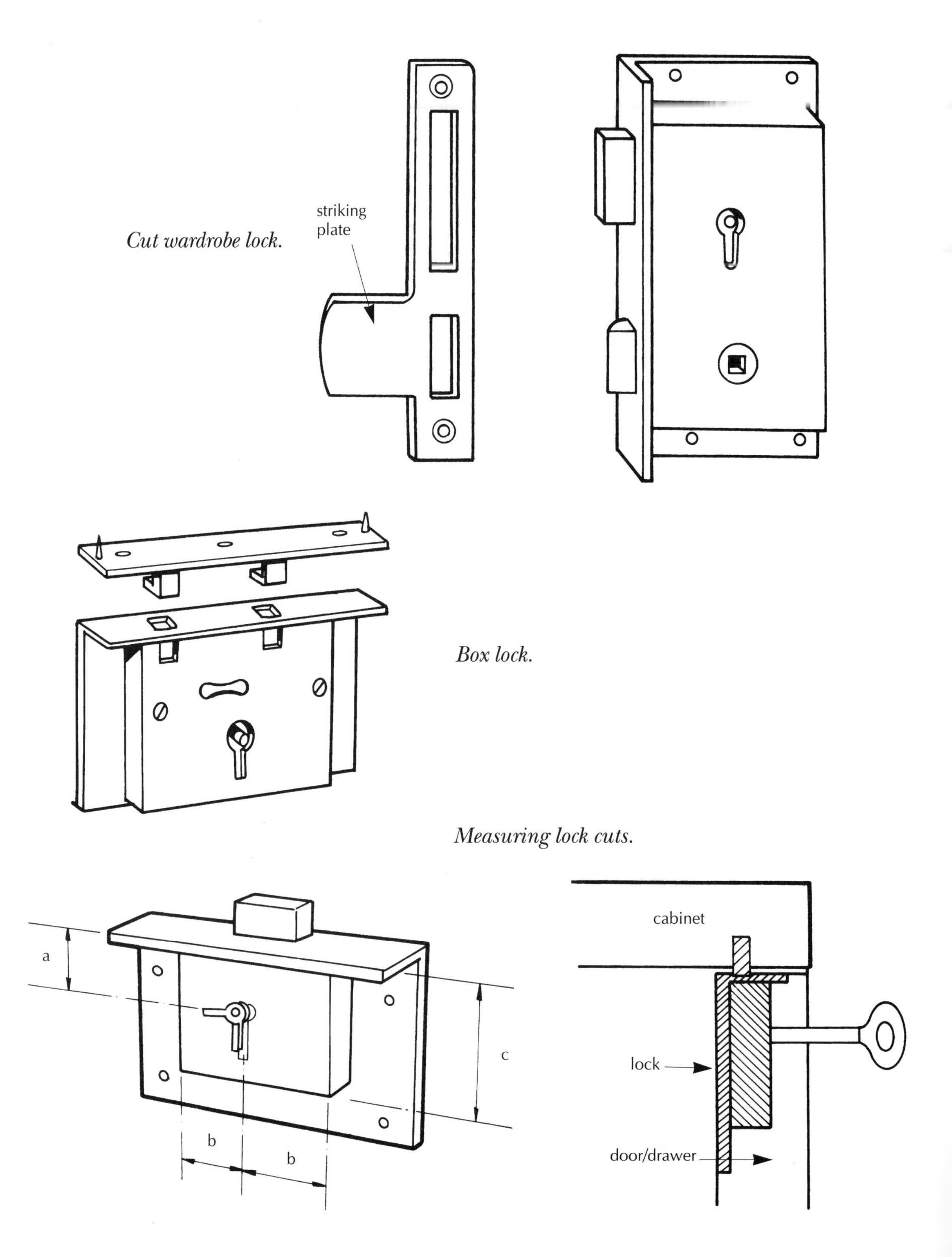

Cut wardrobe lock.

Box lock.

Measuring lock cuts.

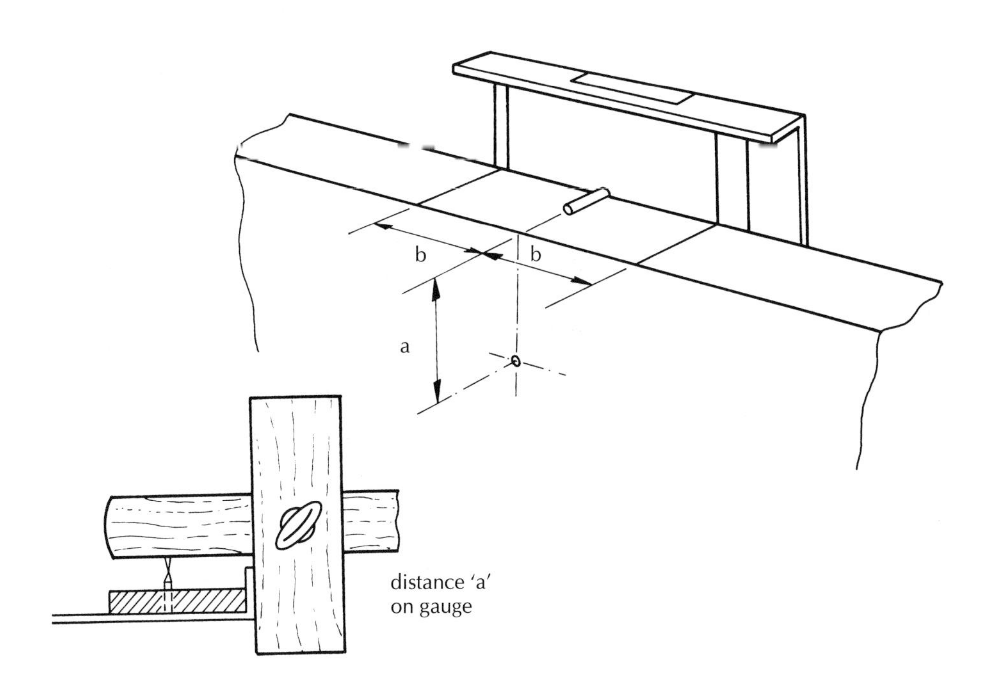

Marking centre lines.

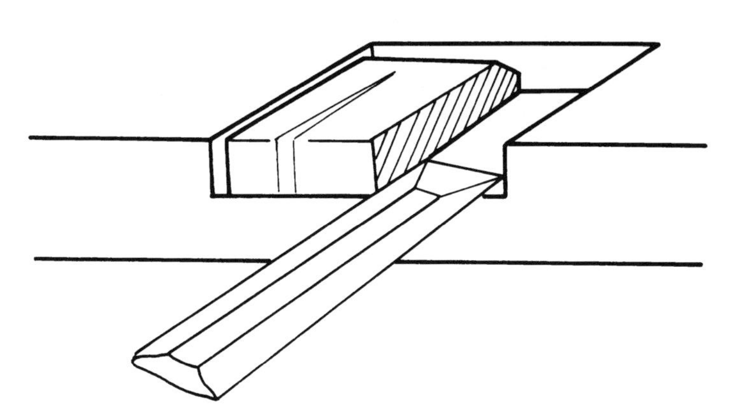

Remove waste with a chisel.

However, once accomplished to a high standard the effect can be pleasantly rewarding.

When preparing for fitting, the first thing that has to be done is to mark out the centre lines for the pin of the lock (a). Do this with a marking gauge. Drill a hole for the pin to enter, taking particular care as it is a common mistake to leave the pin not quite in the centre of the lock. (The hole will have to be enlarged later for the escutcheon to fit.)

Mark out the main part of the lock to the correct specification (b and c). Use a saw to cut and weaken the wood, sawing diagonally, before removing the waste material with a chisel. The next stage is fitting the top lip and back plate of the lock. Slip the lock into position and mark around it with a knife and a cutting gauge. Now chisel to the lines made by the knife and gauge, checking the fit of the lock all the time. Screw the lock loosely into position before removing it again to fit the escutcheon.

Mark around the backplate.

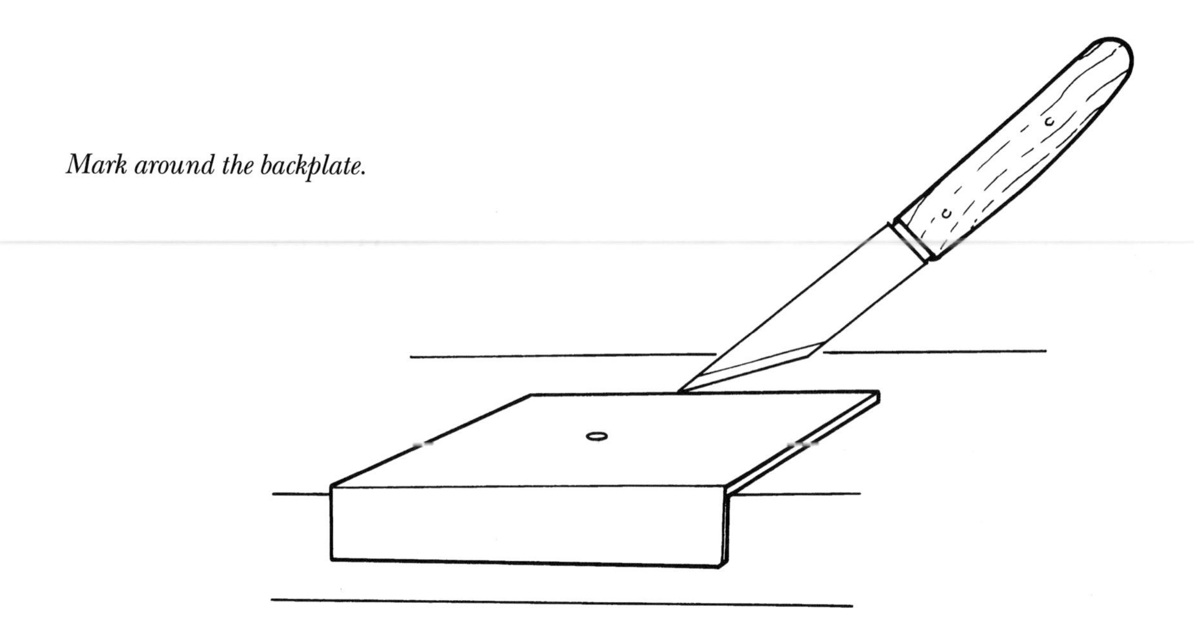

Chisel to knife lines.

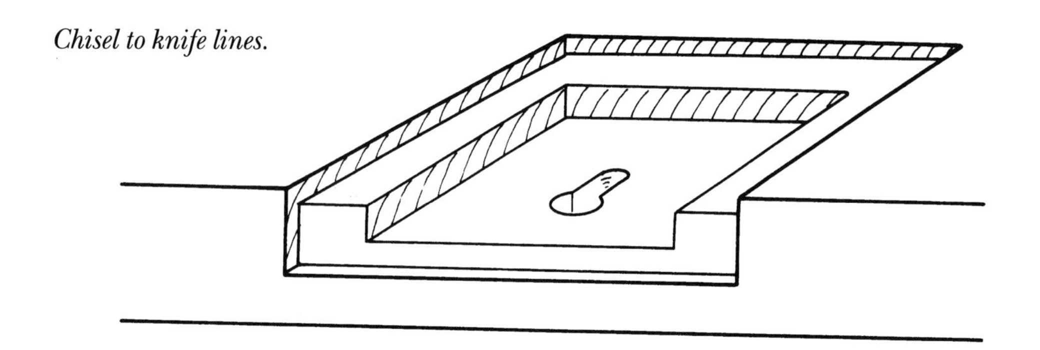

Screw loosely into position.

ESCUTCHEONS

The escutcheon is the key plate that performs the important role of protecting the ground work against damage caused by the constant turning of the key in the lock, while at the same time remaining a decorative embellishment. Since the purpose of the escutcheon is to protect, it is common for it to be made of a robust material that will withstand considerable scraping. Brass is usually used, with timbers such as boxwood offering an alternative. Escutcheons may be inlaid or applied to the surface with an escutcheon pin.

Pressed escutcheons are generally the cheapest to fit. To mark out the escutcheon hold it in the correct place and carefully mark around it. With a drill selected to the correct size matching the outside diameter of the escutcheon enlarge the original pin hole. Using a coping saw, cut away the remaining waste. Clean the hole with a chisel and fit the escutcheon. It may require some force to achieve this – you can apply pressure gently with a G-cramp or a hammer. Finally, sand flush with 180 grit paper.

To fit an *inlaid escutcheon* first cut in the keyhole. Then position the escutcheon precisely and mark its shape with a knife. Recess the shape to the knife outline using a chisel so that the escutcheon plate lies flush with the surface and glue it into position with epoxy glue. Sand it flush.

To fit a *surface escutcheon,* cut out the keyhole first. Then pin the escutcheon into the correct position.

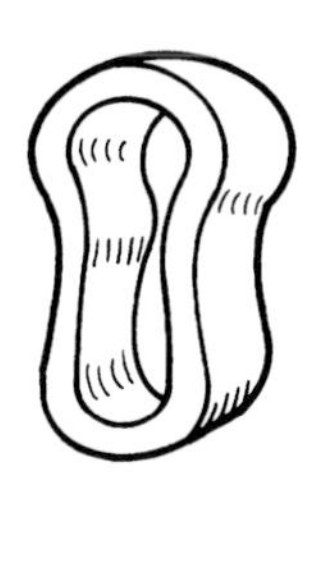

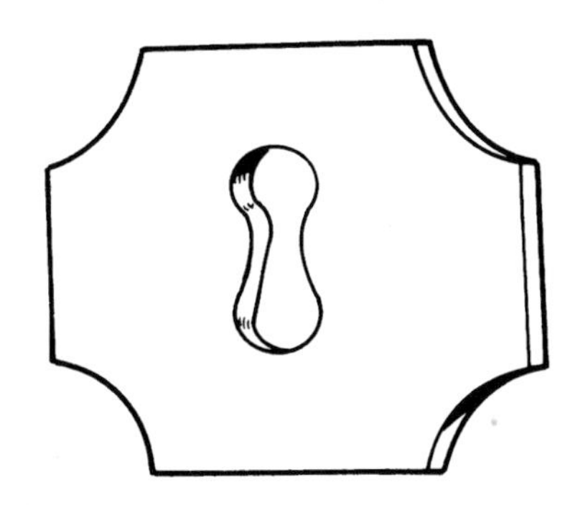

Types of escutcheons.

GLOSSARY

Balance veneer Cheaper veneer used on opposite face or board to balance the face veneer.
Bast Inner bark.
Bear faced Single-shouldered joint.
Burt/Burr Growth at the bottom of the trunk which, when converted to veneer, produces exceptional figure.
Button A means of fixing tops to frames using screws.

Cauls Pieces of wood or metal used to spread the load when pressing veneer onto boards.
Cross bandry Cross grain strips of veneer laid in a decorative manner.

End grain Grain that is exposed across the end of a plank.
Escutchean Metal or wood protection around a key lock.

Face edge The face that is square to the face side.
Face side The chosen face from which all dimensions and marking out are taken from.
Flitches Bundles of veneers.

Green wood Freshly cut, unseasoned wood.

Laminate Strips of wood or veneer glued together.

Lipping Solid wood glued around the edge of man-made board which forms a protective edge.

Muntion A central vertical stile on a door, or a central rail of large drawer bottoms.

Parc To fine-cut timber with a chisel.
Plywood Man-made sheet material made from rotary-peeled constructional veneer glued together.

Runners Wooden strips, often made from teak, upon which a drawer slides.

Sapwood Layer of new growth of wood under the bark.

Tang Forged point on the end of a chisel onto which the handle is fitted.
Template Accurate pattern to help marking out.

Veneer Wood that is cut to a thickness of 0.06mm

Wavy edge Timber plank which still retains its bark.

Wavy grain Irregular patterns in wood fibres.

INDEX

Page numbers in *italics* refer to illustrations.

INDEX